CW00496949

Springer Theses

Recognizing Outstanding Ph.D. Research

Aims and Scope

The series "Springer Theses" brings together a selection of the very best Ph.D. theses from around the world and across the physical sciences. Nominated and endorsed by two recognized specialists, each published volume has been selected for its scientific excellence and the high impact of its contents for the pertinent field of research. For greater accessibility to non-specialists, the published versions include an extended introduction, as well as a foreword by the student's supervisor explaining the special relevance of the work for the field. As a whole, the series will provide a valuable resource both for newcomers to the research fields described, and for other scientists seeking detailed background information on special questions. Finally, it provides an accredited documentation of the valuable contributions made by today's younger generation of scientists.

Theses are accepted into the series by invited nomination only and must fulfill all of the following criteria

- They must be written in good English.
- The topic should fall within the confines of Chemistry, Physics, Earth Sciences, Engineering and related interdisciplinary fields such as Materials, Nanoscience, Chemical Engineering, Complex Systems and Biophysics.
- The work reported in the thesis must represent a significant scientific advance.
- If the thesis includes previously published material, permission to reproduce this must be gained from the respective copyright holder.
- They must have been examined and passed during the 12 months prior to nomination.
- Each thesis should include a foreword by the supervisor outlining the significance of its content.
- The theses should have a clearly defined structure including an introduction accessible to scientists not expert in that particular field.

More information about this series at http://www.springer.com/series/8790

Jaime Ortega Arroyo

Investigation of Nanoscopic Dynamics and Potentials by Interferometric Scattering Microscopy

Doctoral Thesis accepted by
the University of Oxford, Oxford, UK

 Springer

Author
Dr. Jaime Ortega Arroyo
ICFO—The Institute of Photonic
  Sciences
Barcelona
Spain

Supervisor
Prof. Dr. Philipp Kukura
Physical and Theoretical Chemistry
  Laboratory
University of Oxford
Oxford
UK

ISSN 2190-5053   ISSN 2190-5061 (electronic)
Springer Theses
ISBN 978-3-319-77094-9   ISBN 978-3-319-77095-6 (eBook)
https://doi.org/10.1007/978-3-319-77095-6

Library of Congress Control Number: 2018934395

Printed on acid-free paper

This Springer imprint is published by Springer Nature
The registered company is Springer International Publishing AG
The registered company address is: Gewerbestrasse 11, 6330 Cham, Switzerland

I would like to dedicate this thesis to my loving parents.

Supervisor's Foreword

This work describes a breakthrough in optical microscopy in terms of sensitivity and the application of the resulting novel capabilities to a broad range of questions in nanoscience. Microscopy, by definition, is concerned with visualising structure and dynamics on ever decreasing length and timescales. With that comes a need for increasing both temporal resolution and sensitivity with the clear goal of visualising, and thereby studying matter down to the single molecule level. This hurdle was initially taken almost three decades ago in cryogenic environments and has subsequently evolved into an almost standard methodology with far-reaching applications. Prior to the work of Dr. Ortega-Arroyo, optical detection of single molecules has been exclusively limited to resonant detection of chromophores with large absorption cross sections and strong light matter interactions. In this work, Dr. Ortega-Arroyo demonstrates that interferometric scattering microscopy (iSCAT) is capable of detectingand tracking single proteins in solution, and he then applies this unique level of sensitivity to both ultraprecise and rapid single-particle tracking, as well as monitoring self-assembly relevant to the origin of life.

The difficulty associated with detecting single molecules optically without relying on fluorescence comes down to a combination of the discrepancy in physical size of biomolecules, the diffraction limit, and the need to identify the molecule of interest on top of a very large background caused by the environment. The status quo, until this work, has been that it would be impossible to detect single biomolecules with light scattering alone, partially due to the very small signals observed using extinction in previous studies of dye molecules. What Dr. Ortega-Arroyo has shown here, is that, possibly surprisingly, single protein molecules can produce an imaging contrast on the order of a tenth of a percent of the reflected light intensity in an inverted microscope, and that such a signal can be readily detected. The work presented in Chap. 6 I believe, will become an important landmark in optical microscopy, not necessarily because of what was learned about the molecular motor myosin 5a, but because of its implications of what one can study with light.

Dr. Ortega-Arroyo then demonstrates the versatility of scattering, rather than fluorescence-based detection with single-molecule sensitivity. On the one hand, he shows that it can be used to achieve either very high-speed (Chap. 4) or very high precision (Chap. 5) single particle tracking of very small metallic nanoparticle labels. Importantly, it is the improved imaging capabilities that provide key information on how lipids communicate across bilayer membranes and the molecular mechanism behind the remarkable processivity of myosin 5a. On the other hand, Dr. Ortega-Arroyo pushes the sensitivity of the technique even further to directly detects objects as small as single micelles consisting only of a few hundred lipid molecules (Chap. 7) and thereby performs in-situ monitoring of an autocatalytic process relevant to the origin of life and subsequent assembly processes.

This work represents a detailed account of the first label-free detection of single biomolecules in solution and presents a variety of examples of what these capabilities may enable in terms of studying biological and chemical systems. As such, the impact of this work will not be limited to those interested in developing more powerful microscopes, but equally to those who are looking for novel methods to enable measurements that are currently difficult or impossible to do. I am delighted to see it published in the Springer Thesis series.

Oxford, UK   Prof. Dr. Philipp Kukura
January 2018

Abstract

The advent of single-particle tracking and super-resolution imaging techniques has brought forth a revolution in the field of single-molecule optical microscopy. This thesis details the development and subsequent implementation of the technique known as interferometric scattering microscopy as a novel single-molecule tool to study nanoscopic dynamics and their underlying potentials. Specifically, Chap. 2 lays out the theoretical framework and draws comparisons between this technique and other state-of-the-art single-molecule optical approaches. Chapter 3 provides a detailed description for the design and implementation of an interferometric scattering microscope including alignment, instrumentation, hardware interfacing, image processing and respective characterisation to achieve the highest levels of performance. The following two chapters use model systems, namely the diffusion of receptor GM1 in a supported lipid bilayer and the movement of molecular motor myosin 5a, to demonstrate the intrinsic shot-noise-limited nature of the technique, its ability to decouple the temporal resolution from localisation precision, and highlight the importance of taking both parameters into consideration when drawing conclusions about the dynamics of each model system. Chapter 6 provides a proof-of-concept study on the limits of sensitivity and demonstrates for the first time the all-optical label-free imaging, detection and tracking of a single protein. In the last chapter, interferometric scattering microscopy is used to quantitatively study dynamic heterogeneous systems in situ at the single-particle level and thus serves as a proof of principle for future label-free studies beyond the realms of biophysics.

Parts of this thesis have been published in the following journal articles:

Chapter 2:

- Ortega Arroyo, J. & Kukura, P. Interferometric scattering microscopy (iSCAT): new frontiers in ultrafast and ultrasensitive optical microscopy. Phys. Chem. Chem. Phys. 14, 15625–15636 (2012).
- Ortega Arroyo, J. & Kukura, P. Non-fluorescent schemes for single-molecule detection, imaging and spectroscopy. Nat. Photon. 10, 11–17 (2015).

Chapter 3:

- Ortega Arroyo, J., Cole, D., & Kukura, P. Interferometric scattering microscopy and its combination with single-molecule fluorescence imaging. Nat. Protoc. 11, 617–633, (2015).

Chapter 4:

- Spillane, K. M.*, Ortega Arroyo, J.*, de Wit, G., Eggeling, C., Ewers, H., Wallace, M.W. & Kukura, P. Interleaflet coupling and molecular pinning causes anomalous diffusion in bilayer membranes. Nano Lett. 14, 5390–5397 (2014).

Chapter 5:

- Andrecka, J.*, Ortega Arroyo, J.*, de Wit, G., Fineberg, A., MacKinnon, L., Young, G., Takagi, Y., Sellers, J.R. & Kukura, P. Structural dynamics of myosin 5 during processive motion revealed by interferometric scattering microscopy. eLife. 4, e05413 (2015).

Chapter 6:

- Ortega Arroyo, J., Andrecka, J., Billington, N., Takagi, Y., Sellers, J. R. & Kukura, P. Label-free, All-optical detection, imaging, and tracking of a single protein. Nano Lett. 14, 2065–2070 (2014).

Chapter 7:

- Ortega Arroyo, J., Bissette, A., Kukura, P. & Fletcher, S. Visualization of the spontaneous emergence of a complex, dynamic, and autocatalytic system. Proc. Natl. Acad. Sci. U.S.A. 113, 11122–11126, (2016).

Acknowledgements

My life as a graduate student, both inside and outside the laboratory environment, has been moulded by many people, whom without their selfless contributions and support would not have made my time in Oxford a truly memorable and rewarding experience. This thesis is both an acknowledgement and a celebration to those wonderful people, whom I am glad to count all as good friends and the funding agency that made it possible: CONACyT.

First of all, I would to thank Prof. Philipp Kukura for going far and beyond the role of a graduate supervisor, and instead taking the role of a friend and a mentor. Throughout the past years, he has provided me with much more than guidance and fun projects, as he created an environment and gave me the liberty to pursue my passion for finding things out. Even at times when projects refused to give in, I admired his confidence to know that "in the end, it all works out". Moreover I consider it being a privilege to have been part of the first generation and witness how the group matured from just three members to more than a dozen nowadays.

Next I want to acknowledge Matz Liebel, who as part of the first generation spent endless hours in the laboratory with me, learning, building optical contraptions, exchanging fruitful ideas of what the future directions of the field should be and, above all, having fun doing experiments as if the pursuit of knowledge and scientific research were a game to be enjoyed, rather than a job that had to be done for academics sake. I would then like to thank Dr. Joanna Andrecka, who earned the pseudonym of Mama Joanna (MJ), for her passionate care for every member of the group and for always looking out for us both academically and personally. I will never forget the thrill we shared in every project we worked together on, especially on the label-free detection of a protein. We made a great team, and Chaps. 5 and 6 in the thesis are a testament to that they are not mine alone–they are ours.

Probably, the most unexpected and most fascinating projects of all, the origin of life (Chap. 7), would not have been possible without the vision and passion of Andrew Bissette and Prof. Stephen Fletcher and lest not forget the synthesis and characterisation of the necessary reagents. Together, we pushed the technique to new frontiers we did not believe was possible, all while having an absolute fantastic time. I would also like to thank both Gabrielle de Wit and Dr. Katelyn Spillane together with David Marshall (MW group), Dr. Oliver Castel (MW group), Prof. Christian Eggeling and Prof. Mark Wallace for the many fruitful collaborations and discussions on membrane biophysics that led to the results presented in Chap. 4.

Although now belonging to a different field, I want to thank Aleksandar Sebesta for sharing those long hours in the laboratory with me, motivating each other while doing very long and unforgiving experiments, having many passionate discussions and above all being a great friend. Furthermore, I want to express my gratitude and my utmost respect to Dr. Alexander Weigel as a source of knowledge of optics and a role model, whose vision of science research I cherish and commend: as a quest for knowledge that should be performed meticulously and be judged by its quality and not by the popularity of the topic—an element that unfortunately plagues today's science research.

I would also like to thank Prof. James R. Sellers (Jim), Dr. Yasuharu Takagi (Harry) and Prof. Keir Neuman for receiving me with arms wide open at the National Institutes of Health and making me feel at home, away from home. I will never forget that first conference in the San Francisco and every single visit to the NIH. Jim's and Harry's support in the myosin project was critical as they provided us with the purified motor protein constructs and more importantly, they had the patience and faith to entrust their work to us.

As whole, working in the Kukura group for the past years has been an incredible experience, and every single member of it has contributed to it regardless of the choice of project. Christoph Schnedermann, Dr. Torsten Wende, Dr. Alex Duarte and Dr. Jongmin Lim, although our projects are quite different, we always shared a fascination for what each other worked on, attempted to understand what is going on and asked questions that have lead to very interesting developments. I look forward to the many exciting results that are to come from Dan, Gavin and Adam as they have been wonderful students. In addition, it has been a privilege to have such excellent part IIs, especially like Dorcas Tan, my first part II student.

Beyond the academic aspect, I am forever grateful to Anna Jones, for selflessly supporting me day after day and putting up with my long hours in the laboratory and during the write-up. I do not know how the thesis would have been completed without her motivation and help. Similarly, the Aston crew always kept my moral high and helped me keep the fine work-life balance.

A special mention goes to my former supervisor Prof. Edward Grant, who encouraged me to leave my comfort zone in UBC (University of British Columbia) and pointed me into a new direction that ended up being the Kukura group. I will never forget his advise that no matter what you do, as long as you are having fun and are happy, that such choice could not have been wrong, but rather the best possible outcome.

Last but not least, I want to thank my parents, whose everlasting support and complete freedom to choose my path have led me to this very point in time. They may have instilled that bug for finding how things work out, for doing science, but more importantly they allowed me to discover it on my own terms.

The future holds many more challenges and uncertainties; however, I am convinced that the success that is yet to follow will be founded on the experiences and interactions I had with all those people mentioned above.

Contents

List of Figures

Chapter 1
Introduction

Irrespective of the length- and time-scales, the interactions between an object and its surroundings shape the potential energy surface of the system, which in turn determines the type of motion the object exhibits. The search to understand the dynamics of a system, that is to characterise the type of motion and the origin of the interactions that give rise to said behaviour, has brought disciplines as dissimilar as astrophysics and cell biology together. Although the dynamics of a system are most commonly characterised at the ensemble level, the presence of local heterogeneity and thus the likelihood for different interactions pose the scenario that two identical objects may behave very differently. If this is so, a natural question arises: how closely can ensemble measurements capture the dynamics specific to each system, especially considering that local environment changes are known to produce complex behaviour and even regulate the outcome of a process [1]?

The advent of single-molecule approaches has brought upon a revolution in the last few decades across the fields of life sciences and condensed matter physics due to two major reasons [2]. Firstly, measurement of nanoscale phenomena once hidden by ensemble-averaging has not only become feasible but also routine. Secondly, numerous systems simply cannot be investigated by ensemble methods, such as self-assembly processes, mechanisms of motor protein translocation, and complex phase transitions at the nanoscopic scale [3, 4].

Several approaches exist to investigate dynamics at the single-molecule level. One way to categorise them is according to how the single-molecule is detected, i.e. into forced-based and optical-based techniques. Forced-based techniques rely on applying or measuring the change in position due to an applied force. The three most common force-based techniques are optical tweezers [5], magnetic tweezers [6] and atomic force microscopy (AFM) [7]; however other more recent variants like centrifugal-force microscopy [8] have also been developed. All three have distinct advantages and limitations, for instance optical tweezers can access sub-ms temporal resolutions with base-pair resolution (0.34 nm), sufficient to visualise DNA polymerase moving by a single base-pair [9]. Magnetic tweezers cannot only

J. Ortega Arroyo, *Investigation of Nanoscopic Dynamics and Potentials by Interferometric Scattering Microscopy*, Springer Theses, https://doi.org/10.1007/978-3-319-77095-6_1

achieve similar temporal resolution and spatial precision, but also report on rotational forces and torque [10]. However, such high temporal resolution and spatial precision come at the expense of large-sized labels (>500 nm) and limited throughput [11]. Complementary to optical and magnetic tweezers, atomic force microscopy requires no labels and can truly achieve atomic resolution as recently shown by a study where the secondary structure of DNA was resolved [12]. Although high-speed AFM has been achieved experimentally [13], its temporal resolution is far less than that achieved by optical and magnetic tweezers. Furthermore, AFM is still a rather invasive technique, as the interactions between the probe and the sample cannot be fully neglected.

In terms of optical-based approaches, i.e. techniques that rely on light to excite and detect single molecules, the investigation of single-molecule dynamics has been dominated by light microscopy, and thus we will focus exclusively on them in this thesis. The origin of single molecule light microscopy dates back to the seminal experiments by Moerner and Kador, who detected individual molecules at low temperature in a solid by absorption [14]. In its inception, the experiments were performed at very low temperatures to increase the interaction strength between light and matter, which in turn narrowed the line-width due to the exclusive population of the lowest vibrational and rotational levels. This work was quickly followed by another low-temperature experiment by Orrit and coworkers, but in this case relying on fluorescence rather than absorption detection [15]. The far superior signal-to-noise ratio of the latter experiment laid the foundation for the dominance of fluorescence microscopy over all other optical studies at the single molecule level over the next 25 years.

Nowadays, single-molecule fluorescence has become a routine tool to follow the dynamics and underlying potentials within heterogenous systems [16]. Despite having its spatial resolution limited by diffraction, the advent of super-resolution techniques [17] and the demonstration of single nanometre localisation precision [18] has led to a revolution in the life sciences. Nevertheless, there are intrinsic limitations to what fluorescence can achieve. On one hand it suffers from a limited photon flux, and on the other hand its application implies the need for labelling. As a result there are a set of parameters spaces that fluorescence alone cannot access, for instance sub-millisecond temporal resolution with simultaneous sub-10 nm localisation precision for an indefinite observation time. Moreover, there are systems that simply cannot be studied because labelling would indubitably perturb the underlying techniques.

The topic of this thesis is motivated by the aforementioned shortcomings of single-molecule fluorescence, and its goal is to present an alternative approach that can push the temporal resolution, localisation precision and sensitivity domains beyond the realms of fluorescence-based detection. Already experiments by Kusumi et al. have demonstrated that an increase in three orders of magnitude in temporal resolution, thanks to detection of light scattered by gold nanoparticles, has led to further understanding of the underlying dynamics taking place at the plasma membrane of a cell [19]. Furthermore, it is now possible to simultaneously access nanometre precision and sub-millisecond temporal resolution with moderate sized scattering labels (>40 nm in size) using dark-field microscopes [20]. Which leads to the question that embodies the topic of this thesis: can we do better and how far can we go? The answer lies within the exploration of the technique known as interferometric scattering microscopy (iSCAT) [21], both from a technical perspective and from its application to the investigation of nanoscopic dynamics and potentials.

This thesis is arranged as follows:

In Chap. 2, I provide a detailed description of the development and the concept of the technique known as interferometric scattering microscopy in the context of single particle tracking. Special emphasis is given to the different implementations of the technique, mentioning their advantages and disadvantages. A brief mention to the basics of scattering and surface plasmons is covered. Finally, I discuss potential applications and overall limitations of the techniques.

Chapter 3 describes all the instrumentation, hardware interfacing and image processing necessary to achieve the highest levels of performance. The first half of the chapter focuses on the optical design and the hardware components that constitute an iSCAT microscope. Here, the critical aspects of microscope alignment, sample stabilisation, characterisation of the camera properties and hardware triggering are presented. The second half outlines all the image processing tools developed and implemented to convert low quality raw data into shot-noise-limited images that can be analysed using single molecule data analysis algorithms.

Using a solid supported lipid bilayer membrane as a benchmark system, Chap. 4, explores the dynamics of a receptor at the limits of simultaneous localisation precision and time resolution. From a technical perspective, this chapter demonstrates the intrinsic shot-noise limited nature of the technique, its ability to decouple the time resolution from localisation precision, and highlights the importance of taking both parameters into consideration when drawing conclusions. From a membrane biophysics point of view, the source of anomalous diffusion and a mechanism for information exchange across bilayer membranes is discussed.

In Chap. 5, the structural dynamics involved in the translocation of the molecular motor myosin 5a are investigated with unprecedented precision and time resolution. A mechanism of motion based on the constraints imposed by the structure of the protein is proposed in contrast to the full Brownian search model. Characterisation of a spatially constrained and short-lived intermediate state together with the detection of an Angstrom-level structural change provide evidence for the proposed mechanism. Furthermore a new application of iSCAT is discussed based on the use of gold-nanoparticle-protein complexes as signal amplifiers of structural changes.

Chapter 6 describes the label-free imaging capabilities of iSCAT and provides a proof-of-principle experiment for single-protein detection sensitivity. Here, myosin 5a serves as a model system to demonstrate for the first time an all optical approach to simultaneously detect, image, track and therefore study the dynamics of single proteins without labelling.

Chapter 7 departs from the field of biophysics and explores the possibility of using light microscopy to quantitatively study dynamic heterogeneous systems in situ at the single particle level. The self-reproduction of lipid aggregates produced by a bond forming biphasic reaction is addressed. The methods to follow the progress of the reaction and characterise the kinetics are presented together with evidence of complex behaviour present at liquid/liquid and solid/liquid interfaces.

Finally, Chap. 8 concludes the thesis and presents an outlook for future work and the proposal of new single molecule platforms based on the methods and tools described.

References

1. Wennmalm, S., Simon, S.M.: Studying individual events in biology. Annu. Rev. Biochem. **76**, 419–446 (2007)
2. Greenleaf, W.J., Woodside, M.T., Block, S.M.: High-resolution, single-molecule measurements of biomolecular motion. Annu. Rev. Biophys. Biomol. Struct. **36**, 171–190 (2007)
3. Simons, K., Toomre, D.: Lipid rafts and signal transduction. Nat. Rev. Mol. Cell Biol. **1**, 31–39 (2000)
4. Elson, E.L., Fried, E., Dolbow, J.E., Genin, G.M.: Phase separation in biological membranes: integration of theory and experiment. Annu. Rev. Biophys. **39**, 207–226 (2010)
5. Neuman, K.C., Block, S.M.: Optical trapping. Rev. Sci. Instrum. **75**, 2787 (2004)
6. Gosse, C., Croquette, V.: Magnetic tweezers: micromanipulation and force measurement at the molecular level. Biophys. J. **82**, 3314–3329 (2002)
7. Kodera, N., Ando, T.: The path to visualization of walking myosin V by high-speed atomic force microscopy. Biophys. Rev. 1–24 (2014)
8. Halvorsen, K., Wong, W.P.: Massively parallel single-molecule manipulation using centrifugal force. Biophys. J. **98**, L53–L55 (2010)
9. Perkins, T.T.: Ångström-precision optical traps and applications*. Annu. Rev. Biophys. **43**, 279–302 (2014)
10. Lebel, P., Basu, A., Oberstrass, F.C., Tretter, E.M., Bryant, Z.: Gold rotor bead tracking for high-speed measurements of DNA twist, torque and extension. Nat. Methods **11**, 456–462 (2014)
11. Neuman, K.C., Nagy, A.: Single-molecule force spectroscopy: optical tweezers, magnetic tweezers and atomic force microscopy. Nat. Methods **5**, 491–505 (2008)
12. Pyne, A., Thompson, R., Leung, C., Roy, D., Hoogenboom, B.W.: Single-molecule reconstruction of oligonucleotide secondary structure by atomic force microscopy. Small **10**, 3257–3261 (2014)
13. Ando, T., Uchihashi, T., Kodera, N.: High-speed AFM and applications to biomolecular systems. Annu. Rev. Biophys. **42**, 393–414 (2013)
14. Moerner, W.E., Kador, L.: Optical detection and spectroscopy of single molecules in a solid. Phys. Rev. Lett. **62**, 2535–2538 (1989)
15. Orrit, M., Bernard, J.: Single pentacene molecules detected by fluorescence excitation in a p-terphenyl crystal. Phys. Rev. Lett. **65**, 2716–2719 (1990)
16. Cognet, L., Leduc, C., Lounis, B.: Advances in live-cell single-particle tracking and dynamic super-resolution imaging. Curr. Opin. Chem. Biol. **20**, 78–85 (2014)
17. Hell, S.W.: Far-field optical nanoscopy. Science **316**, 1153–1158 (2007)
18. Yildiz, A., et al.: Myosin V walks hand-over-hand: single fluorophore imaging with 1.5-nm localization. Science **300**, 2061–2065 (2003)

19. Kusumi, A., et al.: Paradigm shift of the plasma membrane concept from the two-dimensional continuum fluid to the partitioned fluid: high-speed single-molecule tracking of membrane molecules. Annu. Rev. Biophys. Biomol. Struct. **34**, 351–378 (2005)
20. Ueno, H., et al.: Simple dark-field microscopy with nanometer spatial precision and microsecond temporal resolution. Biophys. J. **98**, 2014–2023 (2010)
21. Lindfors, K., Kalkbrenner, T., Stoller, P., Sandoghdar, V.: Detection and spectroscopy of gold nanoparticles using supercontinuum white light confocal microscopy. Phys. Rev. Lett. **93**, 037401 (2004)

Chapter 2
Non-fluorescent Single-Molecule Approaches to Optical Microscopy

Parts of this chapter have been adapted from the following publication: Ortega Arroyo, J. and Kukura, P. Interferometric scattering microscopy (iSCAT): new frontiers in ultrafast and ultrasensitive optical microscopy. *Phys. Chem. Chem. Phys.* **14**, 15625–15636 (2012). [1] and are copyright (2012) by the Royal Chemistry Society. All work presented in this chapter was performed by myself.

2.1 Introduction

The prevalence of optical spectroscopy in single-molecule experiments has been founded on its ability to extract information about the local environment and underlying dynamics over several time and length-scales by characterising and tuning the properties of optically active molecules and materials. Moreover, the non-invasiveness of optical approaches offer a distinct advantage over surface-sensitive techniques such as atomic force microscopy. Although direct optical detection of single molecules via absorption [2] or extinction spectroscopy [3, 4] is possible, the inherently weak interaction between electromagnetic radiation in the visible regime with matter, due to the gross mismatch between the effective size of a molecule and the wavelength of light (Fig. 2.1), poses significant experimental challenges. These challenges translate into the need to greatly suppress background fluctuations to achieve sufficient signal-to-noise ratios (SNR) for single-molecule detection.

To better understand the sensitivity levels required for direct all-optical single-molecule detection, consider the scenario of an object the size of a molecule placed in the focus of a high numerical aperture objective at room temperature. A typical absorption cross-section for a dye molecule is on the order of 5×10^{-16} cm^2, whereas light can only be focused to an area of 5×10^{-10} cm^2 with a high numerical aperture objective. Under these conditions, the molecule may absorb about one

© Springer International Publishing AG 2018
J. Ortega Arroyo, *Investigation of Nanoscopic Dynamics and Potentials by Interferometric Scattering Microscopy*, Springer Theses,
https://doi.org/10.1007/978-3-319-77095-6_2

per every million incident photons; thus producing a parts-per-million differential signal. Although possible, there are significant experimental challenges associated with a parts-per-million sensitivity and these can be attributed to two main causes. On one hand, external fluctuations such as background scattering and sample drift need to be significantly suppressed, which requires specific measuring conditions and sample stabilisation hardware. On the other hand, the hardware used for taking measurements has intrinsic noise levels orders of magnitude higher than a parts-per-million signal. For instance, commercial laser sources rarely output light with a stability better than 10^{-3}; whereas, photon detectors suffer from dark counts and read-out noise that further degrade the signal-to-noise ratio. Furthermore, in a typical confocal illumination volume, there are as many as 10^7 other molecules interacting with the incident light; further compromising the detection sensitivity. As a result, only highly specialised setups under specific conditions can suppress the background fluctuations to a sufficient level to perform direct optical detection.

In contrast to direct optical detection, fluorescence-based single-molecule detection [5] overcomes the challenges associated with a parts-per-million sensitivity by detecting photons shifted to a lower energy compared to the excitation source; thus converting a high-background measurement into a zero-background one. Furthermore, one of the greatest advantages of single-molecule fluorescence, beyond its superior background suppression, is that a wide range of molecular-sized dyes and fluorescent proteins can be co-expressed with a target molecule and serve as labels. This property is especially attractive for studies in highly complex and heterogeneous environments such as cells, as it grants the molecule specificity by virtue of a characteristic fluorescence emission spectra. This effectively turns a problem analogous to finding a needle in a haystack, into one where each needle contains a tracking device. In this respect, the fluorescent label reports the position and information about the local environment of the molecule of interest, upon which the process termed single-particle tracking (SPT) is based. Moreover, with the information from SPT, i.e. the position and the type of motion of the particle, the underlying nanoscopic potentials responsible for the observed dynamics can be extracted [6–10].

2.2   Single-Particle Tracking

Resorting to an example, the concept of single-particle tracking can explained on the basis of understanding the diffusion of a receptor embedded in a membrane (Fig. 2.2a). Here, the receptor is labelled by a marker which produces an optical response on the detector, known as the point spread function (PSF). This optical response closely resembles that of a 2D Gaussian function with a full-width-at-half-maximum (FWHM) of roughly half the wavelength of light as dictated by the diffraction limit, more formally referred to as the optical resolution of the system. As the receptor diffuses, the position of the PSF changes as a function of time (Fig. 2.2b). Despite the size of the PSF on the order of hundreds of nanometres, the centre of

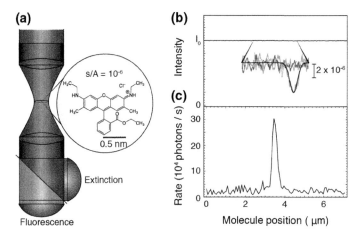

Fig. 2.1 Single-molecule optical microscopy. a Schematic highlighting the relative length-scale mismatch between a dye molecule, such as Rhodamine 6G, and a light source focused to the diffraction limit by a high numerical aperture objective and used to perform single-molecule extinction or fluorescence detection. **b, c** Representative extinction and fluorescence signal magnitudes expected from a single terrylene diimide [3] and a GFP molecule scanned through the focus of an objective; respectively

mass is well defined and thus can be determined with a precision far greater than the resolution of the optical microscope.

Several approaches exist to determine the centre of mass with varying degrees of computational complexity and achievable localisation precision [11, 12]; however for most purposes a fit to a 2D Gaussian function suffices [13]. Repeating this procedure for a sequence of images produces a trajectory with a precision far exceeding the diffraction limit; thus revealing information of the type of motion and possible interactions that lead to said behaviour (Fig. 2.2c). In the absence of external fluctuations in the form of sample drift or vibrations coupled into the system, the signal-to-noise ratio determines the precision and time resolution with which the motion of the label can be followed and thus how closely this motion resembles the dynamics under investigation [14–16]. Therefore if the image quality deteriorates either by shot-noise fluctuations, or appearance of background noise such as out-of-focus fluorescence signal contributions, auto-fluorescence background, or simply by the intrinsic read-out noise of the camera; the centre of mass of the PSF will become less well defined and the corresponding trajectory increasingly noisier, eventually leading to the loss of most of the information content (Fig. 2.2c).

In principle, a greater SNR can be gained by either the collection of more photons and/or the reduction of the background noise. Under the assumption that the background noise is constant, the localisation precision or temporal resolution can be improved at will, either by increasing the light-matter interaction strength or simply incrementing the exposure time. Nevertheless single-particle tracking based on single-emitter fluorescence suffers from two major limitations.

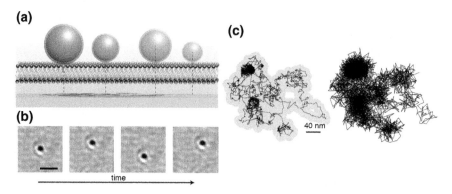

Fig. 2.2 Basics of single-particle tracking. a Illustration of a typical single-particle tracking assay on a membrane bilayer with receptors depicted in red and a spherical marker in bright orange. The trajectory of the labelled receptor as it diffuses across the membrane is represented as the solid red line projected on the glass surface. **b** For labels smaller than the diffraction limit, the PSF resembles a 2D Gaussian. Scale bar: 1 µm **c** Reconstructed trajectory of the motion of the particle PSF in **b**. The true trajectory of the receptor, black line, can be reconstructed with a greater precision than that obtained by the size of the PSF, thick orange line, by finding the centre of mass of the PSF at each time-point. When the SNR of the image decreases so does the quality of the trajectory

Firstly, single emitters possess intrinsic photo physical and chemical properties that restrict the total yield to about 10^7 photons [17, 18]. The two most notable examples are irreversible and reversible photobleaching, both of which become more likely at higher illumination intensities and are the result of the destruction of the fluorophore through the generation and further excitation of triplet states [19]. The development of non-blinking quantum dots [20] and NV-centres in nanodiamonds [21, 22] aim to solve these effects, although routine incorporation of these labels in biological environments remains a challenge and come at a cost of increasing the label size to several tens of nanometres. Secondly and more importantly, all single quantum emitters have a maximum achievable photon flux, otherwise known as optical saturation [17]. This upper limit originates from the need to populate an excited electronic state, which has a lifetime of few to tens of nanoseconds, before a fluorescent photon can be emitted.

These two limitations have far-reaching consequences, namely the SNR can not be tuned arbitrarily and more importantly, the main requirements to resolve dynamics at the nanoscale are coupled: localisation precision, temporal resolution and observation time. From an experimental point-of-view, these limitations place an upper bound to the excitation rates to roughly 1 MHz, which in combination with the typical losses in current state-of-the-art optical microscopes, provide a rough rule of thumb that relates the achievable localisation precision with the corresponding temporal resolution:

$$\sigma(\text{time, space}) = 1 \text{ nm Hz}^{1/2} \tag{2.1}$$

The implications of this relation together with the limited photon yield are quite straightforward: simultaneous high time-resolution and spatial precision measurements spanning more than a handful of data points are inaccessible with single emitters as verified by state-of-the-art tracking experiments in the last decade [23–28]. For instance, localisation with one nanometre precision is possible but comes at the expense of a temporal resolution of a second. Meanwhile, an improvement in the temporal resolution by two orders of magnitude reduces the localisation precision by one order of magnitude.

If the relevant nanoscale dynamics occurred on the time-scale of seconds or slower, the effects of spatial and temporal coupling would not pose a significant problem. Unfortunately, the dynamics of single proteins, lipids and other nanoscale processes occur on time-scales at least three orders of magnitude faster. Returning to the membrane biophysics example, a single lipid in a supported lipid bilayer typically diffuses at a rate of $1\,\mu m^2 s^{-1}$. This is equivalent to stating that within one microsecond, the lipid moves by two nanometres as estimated from the root mean square displacement [29]. Yet, this scenario is quite ideal as motion in a bilayer membrane is considerably slowed down by the reduced dimensionality and the higher viscosity provided by the lipid environment, when compared to the Brownian motion of a ten nanometre protein in solution. In the latter case, the protein would move by a distance equivalent to its own size within a microsecond. Thus, under these conditions the use of single-emitter-based fluorescence to investigate nanoscopic dynamics is analogous to filming a flying bullet with a mechanically shuttered camera from the early twentieth century.

Furthermore most biological molecules do not fluoresce and thus require either chemical or genetic labelling which raises two important issues. On one hand, labelling efficiency can lead to statistical artefacts and in some cases hide potentially rare events; on the other hand, the label can perturb the properties and underlying dynamics of the system we wish to probe. The latter effect is particularly troublesome when studying systems that are comparable in size to the fluorescent labels themselves. Take for example lipids, where certain dye molecules are known to alter the lipid partitioning and distribution within a membrane, [30, 31] or larger sized scattering labels, [32] which can induce variations in the mobility of lipids. Similarly, processes involving assembly or disassembly on the nanoscale, such as amyloid formation (α-synuclein), actin and tubulin polymerisation/depolymerisation are difficult to monitor with true single molecule resolution due to the necessity of distinguishing bound molecules from those at high concentration in solution. Consequently, despite its widespread use and continuous development, single-emitter-based SPT is unsuitable for the precise study of fast dynamics or processes that span multiple orders of magnitude in time.

2.3   Scattering Detection: An Alternative to Fluorescence

Given the limitations of single-emitter-based microscopy and the discrepancy between reality and expectations, there have been numerous attempts to move beyond fluorescence as the contrast mechanism to study single molecules. One such approach relies on the detection of scattered light, which is not subject to the effects of optical saturation, photobleaching or blinking, and does not require a strong transition dipole. Unlike fluorescence, scattering from a nano-object is analogous to having a "nano-mirror", whereby the photon flux is only limited by the amount of incident power on the nano-object and is determined by the scattering cross-section of the object; thus allowing indefinite observation times. Based on these arguments, scattering-based detection schemes become very attractive to study nanoscopic dynamics thanks to the unlimited supply of photons.

Two main approaches to scattering-detection have emerged over the past decades: purely scattering-based and interferometric-based. To understand the basic differences it is instructive to compare both techniques according to their experimental implementations. For ease of comparison, we do so with respect to a standard epi-fluorescence microscope (Fig. 2.3). Detection of pure scattering, commonly known as dark-field microscopy, relies on the rejection of the illumination light from the detection channel with maximal efficiency to reveal only scattered light, represented in the Fig. 2.3 as placing the detector perpendicular to the incident light. However, in a biologically relevant setting, dark-field illumination is usually achieved by a mismatch in the excitation and detection numerical aperture, or by taking advantage of total-internal-reflection. In scattering interferometry rather than rejecting the background light, it serves as a reference and thereby it is collected and recombined with the scattered light by a either a beam splitter or an additional objective, resulting in reflective and transmission geometries; respectively.

Regardless of the approach, the light irradiating the detector, I_d, can be expressed as:

$$I_d = \frac{1}{2}c\varepsilon |\mathbf{E_r} + \mathbf{E_s}|^2 \tag{2.2}$$

where c represents the speed of light, ε the permittivity, and $\mathbf{E_r}$ and $\mathbf{E_s}$ refer to the reference and scattered electric fields, respectively.

Rather than resorting to an electric field description, it becomes more instructive to work in terms of the number of photons that reach the detector, N, given a certain integration time, a detection efficiency, and an irradiance. Applying this transformation to Eq. (2.2) and expanding the electric field components under the assumption of interference results in:

$$N = N_i \left[\zeta^2 + |s|^2 + 2\zeta |s| \cos \Delta\phi \right], \tag{2.3}$$

where s refers to the complex scattering amplitude, $\Delta\phi$ the phase difference between the scattered and reference electric fields and N_i, the incident number of photons. Depending on the approach, ζ represents either the amplitude of any non-suppressed

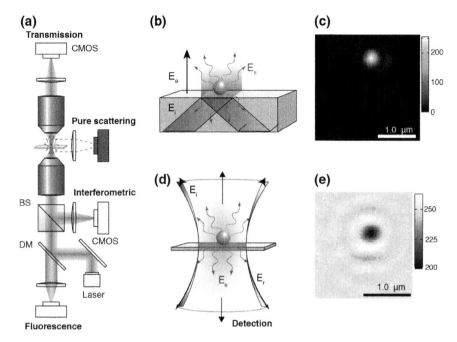

Fig. 2.3 Different implementations of scattering-based detection. a Schematic illustrating the concept behind dark-field and interferometric-based scattering detection with respect to conventional epi-fluorescence microscope. **b** Signal generation for total-internal-reflection darkfield-microscopy: the objective collects the light scattered by the particles, E_s, within the evanescent field produced by total-internal-reflection. **c** Fake colour image of a 40 nm gold particle obtained by TIRDFM. [33] The particle appears as a bright spot over a dark background. Intensity scale in arbitrary units. **d** Signal generation for interferometric scattering microscopy: the objective tightly focuses the incident light, E_i, onto the sample and collects the light reflected, E_r, at the glass cover slip interface and the light backscattered by particles within the illumination region, E_s. **e** Analogous image of a 40 nm gold particle detected by iSCAT. The particle now appears as a dark spot over a constant bright background. Intensity scale in arbitrary units

background, the transmissivity or the reflectivity of the interface. For dark-field microscopy, the detected signal is dominated by the purely scattering term, $|s|^2$, while the remaining terms are negligible for sufficiently small scatterers. Unlike dark-field, in scattering interferometry the reference contribution, ζ^2, dominates the light intensity at the detector and serves as a high-signal baseline containing no information about the sample. Furthermore, as long as the optical path difference between E_r and E_s is smaller than the coherence length of the illumination source, the electric fields will interfere and make the third term in Eq. 2.3 non-negligible.

The relative magnitude between the reference and scattering amplitudes produces two general cases for interferometric detection. On one hand, when the pure and interference terms compete ($\zeta \leq s$), the benefits of interferometry are lost and in fact the measurement becomes analogous to a dark-field one with high background levels [34–36]. On the other hand, in the limit of a very weak scatterer ($\zeta \gg s$) the purely scattering term becomes negligible, and the detected signal can be expressed as:

$$N \approx N_i \left[\zeta^2 + 2\zeta |s| \cos \Delta\phi \right]. \tag{2.4}$$

The immediate consequence of Eq. 2.4 is that the interference term becomes the only contribution that contains information from the sample. Therefore, the signal produced by a weak scatterer is represented by a small change on top of a large background, which we denote as the contrast. This is a major difference compared to both dark-field and fluorescence approaches, where the measurement is ideally performed over a zero-background one. After algebraic manipulation, the contrast can expressed as:

$$\text{Contrast} \approx 2 \, \frac{|s| \cos \Delta\phi}{\zeta}. \tag{2.5}$$

The above equation has enormous implications, namely, the size of the signal is fixed by: the properties of the scatterer, its local environment and the reference amplitude. In other words the contrast is independent of the photon flux and serves as fingerprint for each scatterer [36]. Furthermore, this expression provides a simple means for tuning the contrast, i.e. by controlling the reference amplitude [37].

In most scenarios, the reference term is constant and determined either by the transmissivity or reflectivity of the sample. Therefore the noise of the system, under the most optimised experimental settings, is determined by the fluctuations in the background induced by shot noise. The magnitude in the relative signal fluctuations scale with the number of detected photons as $N^{-1/2}$. Under this consideration the signal-to-noise ratio can be written as:

$$\text{SNR} \propto \text{Contrast}\sqrt{N}. \tag{2.6}$$

A similar expression (SNR $\propto \sqrt{N}$) can be equally derived for dark-field under shot noise conditions. Irrespective of the approach, the key feature of scattering-detection is that, unlike fluorescence, the detected photon flux is only limited by the amount of light incident on the sample. This implies that an arbitrary temporal resolution, localisation precision and sensitivity can be achieved by just tuning the amount of light incident on the sample. This is a very strong statement, yet should not be surprising for a shot noise-limited experiment.

Based on these claims, both scattering approaches seem to greatly surpass the SNR performance of fluorescence detection in single-particle-tracking. Furthermore under shot noise-limited conditions, both dark-field and scattering interferometry attain similar results. So this raises the following two questions: firstly, how feasible and under what conditions is it possible to perform single-particle-tracking experiments in the shot-noise limit? Secondly, are there any significant advantages of interferometric detection over dark-field or vice-versa?

To address the above questions, one must consider the effect of non-negligible background scattering to the SNR of an experiment. As an example, consider the amount of light scattered by a relatively large, 160 nm silica bead with a scattering cross-section in water on the order of 3×10^{-12} cm^2. Compared to the typical diffraction-limited area of a focused beam, only one in a hundred incident photons are scattered by the particle. Although this number is four orders of magnitude larger than for a dye-sized molecule, achieving a localisation precision on the order of one nanometre requires a SNR between 30–60, [15, 16, 23] meaning that any spurious reflections have to be eliminated to at least a level of 0.01% relative to the incident light.

The difficulty associated with this task can be understood by considering that a standard high numerical aperture microscope objective consists of several lenses, each of which will reflect at least 0.1% of the incident light, even in the presence of high-quality anti-reflection coatings. The presence of spurious reflections further deteriorate the achievable SNR, because their signal contribution also scale linearly with the incident light. Some typical sources of these spurious reflections are: dust particles, inclusions in microscope immersion oil, surface inhomogeneities and refractive index changes.

In the case of interferometry, these spurious reflections do not pose such a problem as they become typically overwhelmed by the reference source. In addition, because these spurious contributions are constant, they can be easily subtracted without compromising the image quality. In principle, such a subtraction is also possible in dark-field yet the limited dynamic range found in most detectors compromises its performance. This mostly results from the large spread in signal intensities measured on top of a low background for dark-field; in contrast to the small signal variations measured on top of a large and constant baseline for interferometric scattering.

Although the effect of the label size on single particle tracking experiments is not precisely known, it is highly desirable to use smaller-sized labels than a 160 nm silica bead. This is especially true in the life sciences, where most of the biological nanomachinery ranges in size from a few to several tens of nanometres. It is then advantageous to find smaller-sized labels that produce the same scattered photon flux as their larger-sized counterparts. Metal nano-particles, specifically gold and silver, exhibit strong plasmon resonances in the visible regime (Fig. 2.4a, b). These plasmon resonances are the product of collective oscillations in the electron cloud caused by the interaction of the metallic particle with an oscillating electric field. These collective electron cloud oscillations in turn lead to an enhancement in the scattering and absorption cross-sections compared to other molecules. In fact, a 40 nm gold nano-particle, AuNP, illuminated at 532 nm scatters approximately the same amount of photons in water as a 160 nm silica bead.

To precisely understand the size-dependence of the signal in each approach, one must consider the factors involved in the complexed-value scattering amplitude $s = |s| \exp(i\phi_s)$, which for objects with diameters $D \ll \lambda$ is determined by:

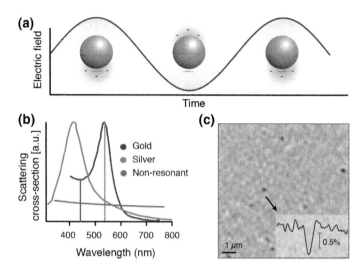

Fig. 2.4 Basics of localised surface plasmon resonance. a Scheme depicting the interaction of a metal nano-particle with an oscillating electromagnetic field, which results in localised oscillations in the electron-cloud density of the particle; otherwise known as localised surface plasmon resonance. **b** Wavelength dependence of the scattering cross-section for different nano-particles. **c** Image of 10 nm Au particles spin-coated onto the surface of a clean glass cover slip immersed in water obtained by illumination at $\lambda = 532$ nm with iSCAT microscopy. Variations in the background correspond to scattering signal contributions from tiny differences in the glass surface. A cross-section of the image is shown to indicate the extinction in signal produced by a single Au particle compared to the glass surface roughness

$$s = \gamma\, \epsilon_{\mathrm{m}}(\lambda)\pi\, \frac{D^3}{2\lambda^2} \frac{\epsilon_{\mathrm{p}}(\lambda) - \epsilon_{\mathrm{m}}(\lambda)}{\epsilon_{\mathrm{p}}(\lambda) + 2\epsilon_{\mathrm{m}}(\lambda)}. \qquad (2.7)$$

Here, γ denotes a proportionality constant, and $\epsilon_{\mathrm{m/p}}$ refers to the complex-valued dielectric constant of the medium and particle, respectively. With this expression, the size dependence on the detected scattering signal for each technique is evident: D^6 versus D^3 for dark-field and interferometric-based approaches, respectively. This reduction in size-dependence for interferometric approaches, means that the signal scales linearly with particle volume. As a result any technique that uses the principle of interferometric detection, has the advantage that reducing the particle size by a factor of 2 carries with it an almost order of magnitude smaller drop in the detected signal (from 64 to 8) compared to direct scattering detection approaches. In the extreme case of sensitivity, particles the size of proteins ($D = 5$–10 nm) can be detected [38, 39].

Close inspection of Eq. 2.7 further reveals that both approaches are not background-free, as any species within the sample will contribute to the detected scattered light as long as $\epsilon_{\mathrm{p}} \neq \epsilon_{\mathrm{m}}$; thus, leading to a lack of signal specificity. For instance, surface variations at the nanoscale such as substrate roughness can be modelled as tiny, randomly shaped nanoparticles made of glass sitting on top of an atomically flat glass surface. Since their refractive index differs from that of the surrounding water, these surface irregularities produce a scattering signal. In fact, exactly this type of behaviour is observed in Fig. 2.4c. This lack of specificity can be viewed as an advantage rather than a short-coming over single-molecule fluorescence as it opens the possibility to perform label-free single-molecule optical microscopy.

2.4 Interferometric Scattering

Today, there are several approaches that take advantage of interferometry more or less directly. They include extinction-based methods [40, 41] photo-thermal detection [42, 43] and techniques using external [44, 45] or common-path referencing [34]. Photo-thermal detection has the unique advantage of enabling the identification of resonant objects even in strongly scattering environments, while externally referenced interferometric detection allows fine-tuning of the phase difference, $\Delta\phi$, and thus the signal magnitude. Common-path interferometric scattering detection dates back to 1965 with the technique known as interference reflection microscopy (IRM) developed by Curtis [46] and further optimised by Ploem [47] with the introduction of the anti-flex illumination scheme. As a precursor to iSCAT, this detection technique was used to study cell adhesion [48–50], thicknesses of thin films over glass substrates [51], interactions between large colloidal particles and surfaces [52], reconstruction of interfacial topology [53, 54], and phase separation in lipid bilayers [51]. The main differences between these techniques with iSCAT is that the latter uses a coherent light source to increase the contrast [50].

In its most recent incarnation, common-path interferometric scattering detection has the major advantage of phase stability and operating in wide-field mode just like fluorescence microscopy. Large field of views on the order of 100s of μm^2 and recording speeds approaching the μs frontier with nm localisation precision are possible. At the same time, the associated experimental setup is comparatively trivial. In fact, most current confocal microscopes are already performing iSCAT experiments everyday; they simply do not have a detector at the appropriate position to record the relevant signal.

A particularly attractive implementation of iSCAT for the life science uses the reflection from the glass cover slip and imaging medium interface as the reference source to achieve homodyne detection. The low reflectivity of the interface (0.5% for glass-water) ensures that most of the incident light is transmitted by the sample, while the scattered light from small particles is often isotropic. As a result the relative amplitudes of scattered and reflected light fields are comparable, which increases the signal contrast of the scatterer. The principle of the technique therefore relies on the illumination and subsequent collection of the light that returns from the sample. Separation of the illumination and the detection channels is achieved by a beam splitter, much a like a dichroic mirror in fluorescence microscopy, except that there is no chromatic difference between the two beams. There are several variants of this technique based upon the type of the illumination and detection employed, each of which presents unique advantages and disadvantages (Fig. 2.5).

2.4.1 Confocal Detection

In its first implementation, iSCAT was performed by using confocal illumination and a point-detector with the advantage to spatially select the scattering signals with a confocal pinhole (Fig. 2.5a). Images were acquired by raster scanning the sample across the focused beam and recording the reflected light intensity as a function of sample position. Despite the fact that this approach is capable of detecting gold nanoparticles down to 5 nm, [34, 35] there are three major drawbacks. Firstly, illumination intensity fluctuations translate into signal variations in the image as the sample is scanned. This is quite problematic as commercially available laser systems rarely stabilise the output intensities to better than 0.05%; thereby, limiting the sensitivity. Secondly, total-internal-reflection from the sample, which needs to be removed or avoided, restricts the numerical aperture below 1.3 and 1.0 for water and air, respectively. Thirdly, the image acquisition speed is intrinsically slow, on the order of seconds; thus making the observation of any fast dynamics impossible.

The key to minimising the effect of all of the above limitations is to switch to either a Koehler or a beam scanning illumination, and from point to wide-field detection [55].

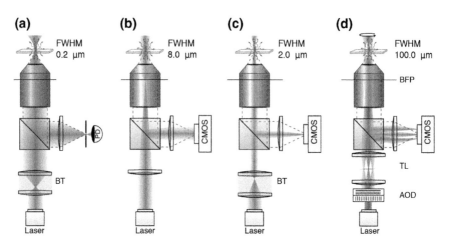

Fig. 2.5 Variants of interferometric scattering setups. Experimental setup diagrams for: **a** confocal detection mode using a point-like detector and a piezo translation stage to raster scan the sample; **b–c** non-scanned wide-field detection using Koehler and pencil-like illumination; **d** confocal beam scanning illumination using acousto-optic deflectors arranged in a telecentric relay system. Dotted lines: scattered light, BFP: back-focal plane of the objective, PD: photodiode, BS: beam-splitter, AOD: acousto-optic deflector, TL: telecentric lens system, BT: beam telescope, CMOS: camera

2.4.2 Non-scanned Wide-Field Detection

The basic operating principle of iSCAT only requires the detection of scattered and reflected light from the sample. In analogy to epi-fluorescence microscopy, this can be achieved by Koehler illumination and subsequent imaging onto a digital camera. The raw scattering image obtained in this fashion is dominated by the reflected illumination beam, with a full width at half maximum determined both by the size of the incident collimated beam and the focal length used to image the light source onto the back focal plane of the objective. In terms of the scattering contrast, the most important parameter is the ratio of scattering amplitude to the size of the diffraction-limited beam. Analogous to fluorescence, the excitation spot size is irrelevant for the final resolution obtained on a camera. After all, the pixels on the CCD camera do not have any knowledge over how the sample was illuminated, but rather record a diffraction limited image of the sample region on their pixels.

The major difference of the Koehler illumination scheme with respect to confocal is the placement of a lens, termed the wide-field lens, along the illumination path, which focuses the beam into the back focal plane of the microscope objective. As a result of this, the objective collimates the incident beam at the sample. The light back-reflected from the glass-water interface together with the back-scattered light from the sample are then collected by the objective and sent into the detection arm of the setup by partially reflecting off the beam splitter. Finally, an image is projected onto a CMOS camera with a lens placed conjugate to the back focal plane of the objective, thus ensuring that the scattered light is focused while the reflected one is collimated.

The quality of the image for Koehler iSCAT depends mainly on three inherent properties of the illumination source: mode stability, Gaussian-like mode character and coherence length. The first two aspects can be controlled either by spatially filtering the output of a laser using a pinhole or by coupling it into a single-mode fibre. In this way mode instabilities are translated into intensity fluctuations, which are accounted for with normalisation of the images by the average pixel value. In regards to the coherence length, iSCAT requires the coherence length to be longer than the interaction distance between the interface and the scatterer in order for the scattered and reflected fields to interfere. However, an increase in the spatial coherence introduces imaging artefacts that significantly corrupt the image quality [56, 57]. The most common of these artefacts are speckle noise and unwanted interference patterns originating from back-reflections. Unfortunately the multiple optical elements present in a microscope objective constitute the main source of back-reflections in an iSCAT microscope [35]. The intensity of these interference patterns, and therefore image degradation, correlates with how tightly the illumination source is focused into the back focal plane of the objective. This would not pose a problem if these modulations remained constant during a typical image acquisition sequence; but such would require an interferometrically stable microscope! This inherently restricts the achievable field of view without discernible image corruption to about $8 \times 8\,\mu m^2$.

Alternatively, if a larger field of view is desired, a dramatic increase in experimental complexity is required to achieve interferometric stability.

A close variant of this approach is to under-fill the objective with a collimated beam [58] as opposed to focusing into the back focal plane. Such a scheme produces a pencil-like illumination profile with an area on the order of 2 μm full-width-at-half-maximum (Fig. 2.5c). This has two major benefits: in first instance, it avoids total-internal-reflection without any loss in signal contrast; and secondly, it minimises the coherent artefact effect. The reduction in the coherent artefact effect results from the fact that the back-reflections are comparable in size to the interface reflection. There-fore, the interference between these two electric fields only causes slowly varying modulations in the overall intensity, which can be accounted for by normalising the image.

The simplicity, moderate-sized fields-of-view, and stability of non-scanned iSCAT makes it the prime choice for two types of experiments: high-speed single particle tracking [58, 59] and label-free bio-sensing [39].

2.4.3 Confocal Beam Scanning Wide-Field Detection

One of the main drawbacks of non-scanned iSCAT is the reliance on the stability of a Gaussian-like mode and the limited field of view caused by coherent artefacts. Larger fields-of-view with higher image quality compared to Koehler illumination can be obtained by using a variant of confocal illumination: confocal beam scanning [55]. Here, rather than raster scanning the sample, a beam focused to a size of less than a micron in FWHM is periodically scanned across the sample at a fixed frequency that is much faster than the exposure time of the camera (Fig. 2.5d). In practice, two beam deflectors operating at different fixed frequencies and scanning in orthogonal directions are required to achieve this type of illumination. Furthermore the beam deflectors are arranged in a telecentric relay imaging system that is conjugate to the back focal plane of the microscope objective (Fig. 2.5d). As a result the camera records what appears to be an evenly illuminated sample region regardless of the mode output by the laser. Moreover the scanning and implicit averaging that takes place, turn the unwanted interference fringes caused by coherent artefacts into pixel variations that no longer degrade the image, analogous to a lock-in detection.

Despite the larger field of view and image quality, there are two main consider-ations that must be addressed. Firstly, given the same average photon flux on the sample, the peak intensity is higher than in non-scanned illumination, due to the small size of the beam being rapidly displaced. Secondly, the maximum frame rate will be limited by the scanning rate of the beam deflectors; since a certain number of cycles are required to produce a homogeneous illumination.

2.5 Applications

2.5.1 Lateral Single-Particle Tracking

One of the most representative advances in the life-science derived from the advent of single-molecule detection via optical microscopy has been the ability to follow the lateral position of individual components of a largely heterogenous system. This has allowed the discovery and proposal of different models across different biological systems ranging from corralled diffusion in cellular membranes [60] to the stepping behaviour of cytoskeletal motor proteins [26, 61, 62]. As mentioned previously, one of the main limitations of the current approaches to lateral single-particle tracking has been the mismatch between the length- and time-scale of the dynamics under study and those achieved experimentally. This mostly results from the coupling between at least two of the following experimental parameters: temporal resolution, localisation precision, label size and observation period. Unlike fluorescence and purely-scattering approaches, interferometric scattering microscopy, when performed in the shot noise-limited regime, relaxes the coupling between all these experimental parameters, which means that a richer parameter space, inaccessible to other techniques, can now be explored as will be shown with the following examples.

An area of study that greatly benefits from the uncoupling of temporal resolution, localisation precision and label size is membrane biophysics due to the mobility of its components occurring mostly in two dimensions. By following the motion of receptors, membrane proteins or individual lipids, information about the type of motion, and the interactions between different membrane components can be extracted as will be discussed in Chap. 4. As a proof of principle that iSCAT can achieve simultaneous localisation precision and temporal resolution, Fig. 2.6a shows a track from a 20 nm gold particle attached to a single lipid via a streptavidin-biotin linker with 2 nm localisation precision taken at an exposure time of $10 \mu s$. Here the temporal resolution was limited by the camera settings but as long as the measurement is shot noise-limited, the temporal resolution can be increased arbitrarily by simply incrementing the incident photon flux on the sample, as recently demonstrated with a similar measurement achieving MHz frame rates [59].

Although a 20 nm gold particle is considered quite small for scattering-based approaches, fluorescent labels such as fluorescent proteins or quantum dots do not typically exceed 5–6 nm. Previous studies of iSCAT have demonstrated the sensitivity to detect immobilised quantum dots [63] and gold particles of 5 and 10 nm in size [35] and thus the ability to use them as labels for lateral tracking. As one step further in the applicability and the sensitivity of the technique, Fig. 2.6b shows an image sequence from both the scattered and fluorescent channels of the same lipid system as above, but with a quantum dot instead of a gold particle. The identity of the quantum dot was assessed by correlative fluorescence and iSCAT measurements, specifically by the presence of blinking in the fluorescence channel. This result illustrates the advantage of iSCAT over fluorescence where the position of the particle is never lost with respect to purely scattering-based detection. Here the sensitivity was incremented

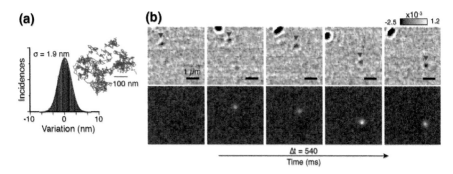

Fig. 2.6 **Capabilities of interferometric scattering microscopy for lateral single-particle tracking. a** Simultaneous high localisation precision and temporal resolution: representative trajectory of a 20 nm AuNP diffusing on a supported lipid bilayer, acquired at 50,000 frames/s and with a nominal localisation precision of 2.0 nm, as determined by the spread in the distances between two immobilised particles. iSCAT illumination: 532 nm. **b** Sensitivity to detect small non-resonant scatterers: correlative iSCAT (top) and fluorescence (bottom) imaging of a single quantum dot (red markers) diffusing on the same supported lipid bilayer system as in **a**. Note that the quantum dot is undetected in the first frame of the fluorescence channel. Sample: supported lipid bilayer constituted by 99.9% DOPE and doped at 0.1% with biotin-labelled DPPE lipid. Streptavidin functionalised quantum dots and AuNPs bind to the DPPE lipids via their streptavidin-biotin moieties. Fluorescence excitation: 473 nm. iSCAT illumination: 630 nm

by temporally averaging frames together rather than increasing the incident photon flux. Both approaches achieve the same result and thus are interchangeable as long as the measurement is shot noise-limited over the time-scale of the frame averaging window.

The previous point does not intend to suggest that iSCAT should substitute fluorescence, instead, it encourages the combination of both techniques to access an additional parameter space. For example, simultaneous iSCAT and fluorescence can be used to show the temporal correlation between the position and the execution of certain events as will later be covered in Chap. 5 with the stepping motion of myosin 5a. Similarly, spatially registered iSCAT and fluorescence measurements, each reporting on a different label, can be exploited to study the orientation of a system with a known structure. Namely, if the physical distance between both labels remains the same, the projection of these two signals on the image plane can be used to find the relative orientation of the labels and the rigid body that connects them; analogous to a GPS system. This precise approach was used to follow the dynamics of the Simian Virus 40 virus-like particles labelled with quantum dots on a supported lipid bilayer [55].

Up to now only spherical particles have been considered in SPT for iSCAT, mostly for the sake of simplicity, since the signal detected is invariant to the orientation of the label. However, the same orientation information described in the previous paragraph can be recovered by using rod-shaped nano-particles. For nanorods, changes in orientation translate into changes in the intensity distribution over different polarisation channels. This concept has been exploited in fluorescence [64] and purely-scattering-based approaches [65–68] to follow orientational changes of the labelled target in addition to lateral displacements. So far, gold nanorods measurements in iSCAT have materialised via confocal detection mode and by illuminating the sample with a light source with a radially or azimuthally polarised donut mode [37, 69]. It is only a matter of time and the choice of a system before the first high temporal resolution and localisation precision measurements with gold nanorods emerge.

Finally, the most subtle advantage that iSCAT possesses, in terms of lateral tracking, is the access to indefinite observation times, irrespective of the temporal resolution and localisation precision. This means that it is possible to indefinitely monitor a single system, with arbitrary localisation precision and temporal resolution, and determine whether or not the observed, potentially sub-ms, dynamics change as a result of other processes that may occur over a much longer time-scale. An example of the above is the change in mobilities of the envelope proteins in an HIV virus following maturation [70]. Here the limiting factor would be the amount of data produced and its subsequent storage, but this can be addressed by processing the images on the fly [71]. For instance the information from a particle in a megapixel image, 1 MB, can be reduced to four 32-bit parameters (time, lateral position and amplitude), which corresponds to just 16 bytes and thus constitutes a reduction of almost five orders of magnitude in data load.

All these results demonstrate many of the fundamental advantages of iSCAT with respect to other single-molecule optical microscopy approaches; namely the ability to observe new dynamics and do so with a sensitivity that is beyond the reach of current state-of-the-art approaches. In addition to the biological insight that this technique may provide, it offers an outlook towards the new parameter space that optical microscopy will soon explore and will eventually become routine in the study of dynamics, much like the current single-particle tracking via the centre of mass.

2.5.2 Axial Localisation via Interferometry

Up to now the discussion has concentrated on lateral tracking; nevertheless motion on the nanoscale seldom occurs in just two dimensions. Obtaining a more faithful picture of the underlying dynamics of a system thus requires, to the very least, extending the localisation principle along the optical axis. In fluorescence, axial localisation has been achieved using the concept of either out of focus imaging [72, 73], multiplane detection [74–76], or PSF engineering [77–82]. In the absence of PSF engineering, the diffraction limited spot is approximately three times wider along the optical axis relative to the focus plane, which correlates with a three times worse axial localisation precision.

Although these approaches can be implemented in iSCAT, this technique intrinsically has the potential to perform axial localisation with equal or better precision than its lateral counterpart [52]. Upon drawing the analogy of iSCAT with a Michelson interferometer, whereby the reference arm corresponds to the reflection at the interface and the scatterer to the test arm, the axial position of the scatterer must therefore

be encoded within the signal contrast (Fig. 2.7a, top panel). This comes as a result
of the difference in optical path lengths, OPL, between each electric field. Rewriting
the signal contrast solely in terms of a trigonometric functional form results in:

$$I_c = I_{max} \cos \Delta\phi(z) \tag{2.8}$$

where $I_{max} = 2|s|/|\zeta|$. The above equation reveals that the phase difference between
both electric fields, $\Delta\phi$, modulates the signal contrast, represented by I_c. This phase
difference can be expressed in terms of an OPL component and an intrinsic phase,
ϕ_0 as follows:

$$\Delta\phi(z) = \frac{4\pi}{\lambda} OPL(z) + \phi_0 \tag{2.9}$$

where $OPL = \sum n_i z_i$. The term n corresponds to the refractive index of the material
situated between the glass interface and the centre of mass of the scatterer. Based
on Eq. 2.9, it seems straightforward to extract the axial position from a sinusoidally
modulated signal, yet there are two main points that must be addressed.

Firstly, a complete signal inversion occurs for axial displacements on the order of
$\lambda/4n_i$, which in a medium composed of water and excitation at 530 nm corresponds
to approximately 100 nm (Fig. 2.7a, bottom panel). This means that a unique mapping
between axial position and contrast can only be guaranteed for a maximum range

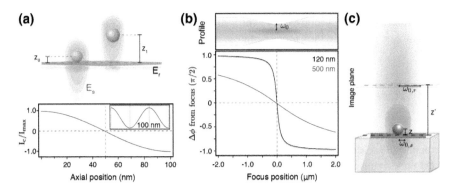

Fig. 2.7 Concept of axial localisation for iSCAT. a The axial position of a scatterer, z_i, depends
on the optical path length difference between the scattered and the reflected electric field. Analogous
to a Michelson interferometer, changes in the optical path length in the test arm result in changes in
phase, which in turn modulate the signal intensity sinusoidally. For illumination at $\lambda = 532$ nm and
for a sample immersed in an aqueous medium, a full signal inversion occurs at an axial displacement
of 100 nm. **b** Concept of the Gouy phase shift for a beam that passes through a point focus. The rate
at which the phase shifts strongly depends on the beam radius at focus, ω_0. **c** Diagram illustrating
the effect of imaging at an out-of-focus position, z', and how this value relates to the axial position
of a scatterer. Note how both the scattered and the reflected electric fields, with their respective
beam radii, accumulate a Gouy phase shift that depends on the value z'. This diagram assumes that
an image is in focus, $z' = 0$, when the illuminating beam is focused on the surface of the sample

of 100 nm. In addition, as the signal inversion occurs, the contrast necessarily must attain a zero value, which itself poses the issue that there exists a set of axial positions that fall below the detection range of a measurement. As a corollary, this makes the technique extremely sensitive to axial displacements and thus ideally-suited for experiments where the motion of the scatterer is constrained within the first few tens of nanometers away from the interface.

Secondly, in an experiment neither the value of maximum contrast, I_{max}, nor the intrinsic phase contributions are defined. An obvious solution involves a calibration with particles fixed at controlled heights, as was recently demonstrated with an experiment involving a more complicated interferometric scheme, originating from a flow cell with a strongly reflective top surface. In this study, charged gold nanoparticles, confined within an electrostatic trap provided by a nanoscopic pocket, were tracked in three dimensions, information which was subsequently used to characterise the underlying trapping potential [8, 83]. The two unknown parameters were calculated from a two-point calibration over an ensemble of particles, where each point of the calibration corresponded to gold particles immobilised to the exterior of a pocket with a known depth [84]. Although this approach serves as a proof of concept, extending such a calibration to a simpler illumination scheme is non-trivial, simply because of the experimental challenge associated with precisely controlling the height of a scatterer without introducing unwanted background scattering.

Moreover this calibration would only work at a precise focus position, given that the intrinsic phase term is composed of a focus dependent phase contribution known as the Gouy phase (ϕ^G). The Gouy phase phenomenon can be explained on the basis of the uncertainty principle, where the action of spatially confining a beam of light in the transverse direction, by focusing, results in an increase in the spread of momenta, which in turn leads to a phase shift [85]. The generality of this phenomenon dictates that the Gouy phase is not unique to microscopes, instead it occurs for any numerical aperture lens. For instance, assuming a point focus, a beam that has passed through the focus acquires a $-\pi$ phase shift relative to a plane or spherical wave travelling the same distance. Although a phase shift of $-\pi/2$ occurs precisely at the focal point, it is of interest to know the rate at which this happens elsewhere along the optical axis. The following equation describes such:

$$\phi_i^G(z') = -\arctan\left(\frac{\lambda}{\pi\,\omega_{0,i}^2}\,z'\right) \qquad (2.10)$$

where ω_0 and z' refer to the $1/e^2$ beam radius at focus and the position of the sample relative to the focus ($z' = 0$), respectively. Close inspection of Eq. 2.10 suggests that the Gouy phase is not only wavelength dependent, but also sensitive to how strongly the beam is focused. Namely, a tightly focused beam will shift phase more rapidly than a weakly focused one (Fig. 2.7b). In addition, the sign convention follows from the assumption that the time-varying factor in the electric field is expressed as $\exp(-i\omega t)$ for a given frequency ω [86, 87].

This discussion so far of the Gouy phase has only referred to the case of transverse confinement of light by focusing, which is analogous to the phase evolution of light diffracted by a small-sized aperture. However this does not cover the scenario of phase evolution from a scatterer, which, following the same analogy, would be akin to that of light diffracted by the complimentary shape of an aperture, i.e. an obstacle. Babinet's principle, where the sum of both complimentary diffracted fields at any point along the optical axis equals the incident field, dictates that a scattering Gouy phase exists, and that it has opposite sign to that caused by focusing [88]. This statement has two subtle points that need to be mentioned when referring to iSCAT. On one hand, in terms of absolute values, the scattering Gouy phase will only ever acquire a maximum phase shift of $\pi/2$ in the far-field compared to π for a focused beam. On the other hand, the beam radius parameter will be different for each field, which in the calculation of the phase difference between scattered and reference field yields:

$$\phi^G(z') = \phi_s^G - \phi_r^G = \arctan\left(\frac{\lambda}{\pi \omega_{0,s}^2}z'\right) - \left[-\pi/2 - \arctan\left(\frac{\lambda}{\pi \omega_{0,r}^2}z'\right)\right] \quad (2.11)$$

After algebraic manipulation using the identity,

$$\arctan(u) + \arctan(v) = \arctan\left(\frac{u+v}{1-uv}\right)$$

and under the approximation that $z' < 100\,\text{nm}$, Eq. 2.11 reduces to:

$$\phi^G(z') = \arctan\left[\frac{\lambda}{\pi}\left(\frac{1}{\omega_{0,s}^2} + \frac{1}{\omega_{0,r}^2}\right)z'\right] + \pi/2 \quad (2.12)$$

The previous derivations can be applied to all forms of iSCAT where a (nearly) focused beam illuminates the sample area (Fig. 2.7c). In the case of Koehler illumination the same results can be extended by considering the next two points. Firstly, the incident beam is collimated at the sample, so the reflected beam undergoes no Gouy phase change. However, due to the Koehler arrangement, the reflected beam gets focused by the objective only to become later collimated by the imaging lenses, thus acquiring a total $-\pi$ phase shift. Secondly, the imaging lens in Koehler iSCAT only focuses the scattered light onto the camera chip, which leads to a $-\pi/2$ phase shift. Taking the difference of these two contributions yields a net shift of $\pi/2$, which reduces Eq. 2.11 to:

$$\phi^G(z') = \arctan\left(\frac{\lambda}{\pi}\frac{1}{\omega_{0,s}^2}z'\right) + \pi/2 \quad (2.13)$$

The last contribution buried within the intrinsic phase refers to the retardation of the electric field upon interacting with a particle. In the limit for small particles, $\lambda \gg D$, where the dipole approximation holds, the retardation phase alludes to whether the induced dipole oscillates, and thereby radiates with or without a phase lag relative to the incident electric field. The scattering phase can be calculated from the complex valued polarisability, α, as follows:

$$\phi_{sca} = \arctan \left(\frac{\text{Im } \alpha}{\text{Re } \alpha} \right) \tag{2.14}$$

The limiting values for the scattering phase correspond to: $\pi/2$ for a particle in resonance with the incident field, and 0 for a perfect dielectric particle or one excited far away from resonance. The expression for the polarisability of a particle is similar to the scattering amplitude shown in Eq. 2.7. In the following equation, for ease of determining the scattering phase, only the contributing terms from the expression of the polarisability of a particle are written:

$$\alpha \propto \frac{n_p^2 - n_m^2}{n_p^2 + 2n_m^2} \tag{2.15}$$

Here the dielectric constant of the particle and the material have been substituted for the respective complex refractive index values, following the relation $\epsilon = n^2$. As an example, the scattering phase of a gold nanoparticle in water illuminated at a $\lambda = 532\,\text{nm}$ can be estimated from the refractive index values of the bulk materials, specifically: $n_{Au} = 0.543 + 2.331i$ and $n_m = 1.335$, as $\phi_{sca} = 0.21\pi$.

Finally with all the phase contributions accounted for, the equation relating the contrast of a scatterer with its axial position and the relative focus of the sample can be written as:

$$I_c(z, z') = I_{max} \cos \left[\frac{4\pi}{\lambda} \text{OPL}(z) + \phi_{sca} + \phi^G(z') \right] \tag{2.16}$$

To validate Eq. 2.16 with physical observations it suffices to analyse the behaviour of the contrast signal when the sample is in focus, and the OPL is negligible. Three general cases occur depending on the dielectric properties of the scatterer with respect to the incident wavelength, specifically whether the particles are either: (i) resonant ($\phi_{sca} \approx \pi/2$), (ii) non-resonant ($\phi_{sca} \approx 0$), or (iii) slightly off-resonant and possess a complex valued refractive index (ϕ_{sca} with values in between).

In the first case, the maximum contrast coincides with the particle being in focus and the resulting signal corresponds to a dip in the baseline, which is interpreted as the product of destructive interference. Moreover as the focus is displaced, so too is the signal contrast and with it the SNR. Conversely, for non-resonant particles, when the sample is in focus (approximately) no signal contrast is expected. However the variability of the Gouy phase with focus can be used to increase the signal for such weak scatterers like viruses or QDs [63]. Metallic nanoparticles produce a non-negligible signal at focus, however, they only reach their maximum contrast at a position slightly out-of-focus, as shown both experimentally [63] and theoretically [86].

Equipped with this understanding of the phase contributions and the commercial availability of feedback-controlled sample stages with nm stability, it should be possible to generate more robust calibration schemes to fully unleash the potential of 3D-iSCAT. Studies so far have shown that iSCAT can deliver not only lateral, but also axial localisation precisions far exceeding the current single-molecule optical microscopy approaches. More importantly, these results highlight the prospective new avenues of research, whereby small labels are used as nanoscopic probes of underlying potentials. So far this approach has been applied to electrostatic traps caused by nanoscale pockets, but extending it to the study of dynamic potentials happening at liquid/liquid interfaces is an endeavour with great promise [89, 90].

2.5.3   Label-Free Imaging

Up to this point, emphasis has been placed on the use of nanoscopic labels to observe the dynamics of a system. Yet, any object will elastically scatter a portion of the incident light, specifically if the refractive index of the material differs from its surrounding environment. This principle can be exploited to perform non-resonant-based imaging, meaning that iSCAT measurements can be extended to any (bio)molecule in the absence of any modifications.

In the context of biomolecules, proteins are an immediate choice for label-free imaging applications due to their ubiquitous occurrence in biological processes, and their significantly larger size compared to lipids and DNA. More often than not, the size of proteins is measured in units of molecular weight rather than nanometres. Therefore it is convenient to show the dependence of the contrast in terms of the molecular weight and refractive index of a protein. One particular feature facilitates this, specifically, that most proteins have on average the same partial specific volume of $v = 0.73 \, \text{cm}^3 \, \text{g}^{-1}$, which is simply the reciprocal of the density [91]. By applying the conversion between molecular weight and volume, $V = M_w / N_a v$, a direct relation between the scattering contrast and the molecular weight of the protein can be obtained:

$$s \propto M_w \frac{n_p^2 - n_m^2}{n_p^2 + 2n_m^2} \tag{2.17}$$

The above relation establishes the following rule of thumb: the iSCAT signal scales linearly with the molecular weight of a protein. To put things in perspective with the discussion of lateral tracking with AuNPs, a 20 nm gold particle has an approximate molecular weight of 50 MDa. Although this number is three orders of magnitude larger than for small proteins such as bovine serum albumin (66.7 kDa), there exist many macromolecular protein complexes with molecular weights in the tens of MDa range. This means that the same SNR can be achieved with these complexes as with gold nanoparticles, and with it, the same temporal resolution and localisation precision. The first piece of evidence for this claim was realised with iSCAT under a Koehler configuration by the detection of immobilised microtubules with a molecular weight per diffraction-limited area of 26 MDa [35]. However no explicit characterisation of the signal contrast or comparison with AuNPs was made in this study.

Experimental confirmation of the above statement resulted from the study of SV40 virions and virus-like particles, which have the same structure as the virions but without the viral genome, on a supported lipid bilayer [92]. In this work, each of these particles was laterally tracked and their signal contrast characterised with respect to 20 nm gold particles. Specifically, the signals for SV40 virions, SV40 virus-like particles and 20 nm AuNPs were: $3.00 \pm 0.87\%$, $1.28 \pm 0.25\%$, and $2.61 \pm 0.65\%$, respectively. Although the exact value of the contrast will depend on the setup configuration, this experiment established a link between the molecular weights of a 20 nm AuNP (50 MDa) and an SV40 virus-like particle (15 MDa). As a result, we can now estimate the contrast of any protein! A follow-up experiment at higher frame rates and localisation precision, performed simultaneously with fluorescence detection, was conducted to study the interactions and underlying dynamics of SV40 virus-like particles with a supported lipid bilayer doped with the target receptor GM1 [55].

Beyond the context of tracking and the detection of (bio)molecules, label-free imaging can be extended to systems that undergo nanoscopic refractive index changes, given the extreme sensitivity of this technique to such variations. Applications exploiting this sensitivity have been numerous and diverse, ranging from the study of cell adhesion, [49, 50, 54] identification of nanoscopic domains in model membranes (to be published), to nanoscale self assembly processes such as the formation of supported lipid bilayers [93].

All in all, these experiments serve as foundations for future experimental endeavours and provide us with a glimpse of what is yet to come. Nevertheless two immediate questions come to mind, specifically: how small a protein can iSCAT detect, and can iSCAT have applications beyond the realms of biophysics? Chapters 6 and 7, aim at providing answers to these questions.

2.6 Conclusion and Outlook

Regardless of the system, this chapter has shown that the ultimate limiting factor for any iSCAT experiment is shot noise, whether it is the quest for high temporal resolution, localisation precision, sensitivity or all three combined. This implies that effectively the photon flux incident on the sample determines the boundaries of any experimental performance. To assess this effect, Fig. 2.8a shows theoretical predictions of localisation precision, for different-sized gold particles, as a function of illumination intensity for an exposure time of $\Delta t = 100 \mu s$ to enable a meaningful comparison with fluorescence. A natural question arises from this statement: at what point does the incident photon flux fail to provide shot noise-limited behaviour in the detection?

In the case of fluorescence, the answer is, when the single emitter reaches optical saturation or photobleaches, whichever happens first. For Fig. 2.8a, only the effects of optical saturation were considered and are represented by the levelling-out of the curve. Although iSCAT does not suffer from these drawbacks, the previous statements hint at the source of these limitations: the interaction of light with the sample, in this instance, the scatterer. Up to now, the effects of the incident photon flux on the properties of the scatterer and its surroundings have been neglected. Considering that gold nanoparticles not only scatter light but also absorb a fraction thereof, localised heating will eventually pose a problem, especially if the interaction with light is resonant [94–96]. In Fig. 2.8b the rise in temperature due to heating of the gold nanoparticle has been calculated based on theoretical models [94]. In addition to thermal effects, gold nanoparticles have been attributed as a source of hot electrons, which depending on the system, can be considered disruptive, as in the case of DNA [97], or advantageous, for example by serving as a local trigger of chemical reactions [98].

Overall, Fig. 2.8 serves as a guide to determine whether the combination of localisation precision, temporal resolution and label size for a certain experiment is feasible. As an example, achieving 1 nm precision with a 20 nm AuNP at an exposure time of 10 μs requires at least an incident intensity of 70 kW/cm^2, which leads to a local temperature rise at the surface of the gold particle of 3 K. In light of these limitations we can ask whether it is possible to do better, and if so, how? One option involves the use of a silver nanoparticles, which require even less incident power due to the higher

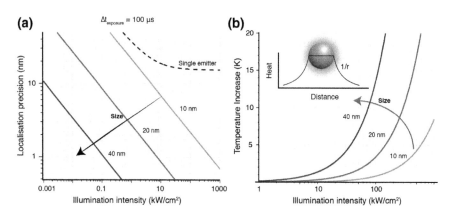

Fig. 2.8 Theoretical limit of localisation precision and temperature considerations. a Localisation precision dependence on incident illumination from different techniques based on current literature for 100 μs exposure time for different-sized gold nanoparticles. **b** Theoretically calculated maximum temperature increase on the surface of a single spherical nano-particle induced by absorption of incident light at 532 nm [94]. Inset: schematic of heat dissipation as a function of distance

scattering cross-section, but come at the expense of an increased absorption cross-section and issues with photo-oxidation. Another approach consists in either using non-resonant particles or illuminating at wavelengths far off resonance, specifically further to the red to minimise photodamage.

This chapter has provided a comprehensive review of iSCAT, both from an experimental and theoretical perspective and has outlined the strengths and weaknesses of the technique. The increase in temporal resolution, whilst keeping localisation precision on the order of nanometers, marks an improvement of more than three orders of magnitude compared to state-of-the-art single-emitter-based techniques. In addition, with the improved sensitivity to perform non-resonant detection, iSCAT promises to become a universal tool for the study of nanoscopic dynamics and potentials.

References

1. Ortega Arroyo, J., Kukura, P.: Interferometric scattering microscopy (iSCAT): new frontiers in ultrafast and ultrasensitive optical microscopy. Phys. Chem. Chem. Phys. **14**, 15625–15636 (2012)
2. Moerner, W.E., Kador, L.: Optical detection and spectroscopy of single molecules in a solid. Phys. Rev. Lett. **62**, 2535–2538 (1989)
3. Kukura, P., Celebrano, M., Renn, A., Sandoghdar, V.: Single-molecule sensitivity in optical absorption at room temperature. J. Phys. Chem. Lett. **1**, 3323–3327 (2010)
4. Chong, S., Min, W., Xie, X.S.: Ground-state depletion microscopy: detection sensitivity of single-molecule optical absorption at room temperature. J. Phys. Chem. Lett. **1**, 3316–3322 (2010)
5. Orrit, M., Bernard, J.: Single pentacene molecules detected by fluorescence excitation in a p-terphenyl crystal. Phys. Rev. Lett. **65**, 2716–2719 (1990)
6. Brokmann, X., Coolen, L., Hermier, J.P., Dahan, M.: Emission properties of single CdSe/ZnS quantum dots close to a dielectric interface. Chem. Phys. **318**, 91–98 (2005)
7. Jin, S., Haggie, P.M., Verkman, A.S.: Single-particle tracking of membrane protein diffusion in a potential: simulation, detection, and application to confined diffusion of CFTR Cl channels. Biophys. J. **93**, 1079–1088 (2007)
8. Krishnan, M., Mojarad, N.M., Kukura, P., Sandoghdar, V.: Geometry-induced electrostatic trapping of nanometric objects in a fluid. Nature **467**, 692–695 (2010)
9. Türkcan, S., et al.: Observing the confinement potential of bacterial pore-forming toxin receptors inside rafts with nonblinking Eu3+-Doped Oxide nanoparticles. Biophys. J. **102**, 2299–2308 (2012)
10. Jaqaman, K., Grinstein, S.: Regulation from within: the cytoskeleton in transmembrane signaling. Trends Cell Biol. **22**, 515–526 (2012)
11. Cheezum, M.K., Walker, W.F., Guilford, W.H.: Quantitative comparison of algorithms for tracking single fluorescent particles. Biophys. J. **81**, 2378–2388 (2001)
12. Sage, D. et al.: Quantitative evaluation of software packages for single-molecule localization microscopy. Nat. Methods (2015)
13. Stallinga, S., Rieger, B.: Accuracy of the gaussian point spread function model in 2D localization microscopy. Opt. Express **18**, 24461–24476 (2010)
14. Bobroff, N.: Position measurement with a resolution and noise limited instrument. Rev. Sci. Instrum. **57**, 1152–1157 (1986)
15. Thompson, R.E., Larson, D.R., Webb, W.W.: Precise nanometer localization analysis for individual fluorescent probes. Biophys. J. **82**, 2775–2783 (2002)

16. Ober, R.J., Ram, S., Ward, E.S.: Localization accuracy in single-molecule microscopy. Biophys. J. **86**, 1185–1200 (2004)
17. Schmidt, T., Schuetz, G.J., Baumgartner, W., Gruber, H.J., Schindler, H.: Characterization of photophysics and mobility of single molecules in a fluid lipid membrane. J. Phys. Chem. **99**, 17662–17668 (1995)
18. Moerner, W.E., Fromm, D.P.: Methods of single-molecule fluorescence spectroscopy and microscopy. Rev. Sci. Instrum. **74**, 3597 (2003)
19. Eggeling, C., Volkmer, A., Seidel, C.A.M.: Molecular photobleaching kinetics of rhodamine 6G by one- and two-photon induced confocal fluorescence microscopy. Chem. Phys. Chem. **6**, 791–804 (2005)
20. Wang, X., et al.: Non-blinking semiconductor nanocrystals. Nature **459**, 686–689 (2009)
21. Fu, C.-C., et al.: Characterization and application of single fluorescent nanodiamonds as cellular biomarkers. Proc. Natl. Acad. Sci. USA **104**, 727–732 (2007)
22. Rabeau, J.R., et al.: Single nitrogen vacancy centers in chemical vapor deposited diamond nanocrystals. Nano Lett. **7**, 3433–3437 (2007)
23. Kubitscheck, U., Kückmann, O., Kues, T., Peters, R.: Imaging and tracking of single GFP molecules in solution. Biophys. J. **78**, 2170–2179 (2000)
24. Kues, T., Peters, R., Kubitscheck, U.: Visualization and tracking of single protein molecules in the cell nucleus. Biophys. J. **80**, 2954–2967 (2001)
25. Seisenberger, G., et al.: Real-time single-molecule imaging of the infection pathway of an adeno-associated virus. Science **294**, 1929–1932 (2001)
26. Yildiz, A., et al.: Myosin V walks hand-over-hand: single fluorophore imaging with 1.5-nm localization. Science **300**, 2061–2065 (2003)
27. Yildiz, A., Selvin, P.R.: Fluorescence imaging with one nanometer accuracy: application to molecular motors. Acc. Chem. Res. **38**, 574–582 (2005)
28. Pierobon, P., et al.: Velocity, processivity, and individual steps of single Myosin V molecules in live cells. Biophys. J. **96**, 4268–4275 (2009)
29. Saxton, M.J., Jacobson, K.: Single-particle tracking: applications to membrane dynamics. Annu. Rev. Biophys. Biomol. Struct. **26**, 373–399 (1997)
30. Kaiser, H.J., et al.: Order of lipid phases in model and plasma membranes. Proc. Natl. Acad. Sci. USA **106**, 16645–16650 (2009)
31. Staneva, G., Seigneuret, M., Conjeaud, H., Puff, N., Angelova, M.I.: Making a tool of an artifact: the application of photoinduced lo domains in giant unilamellar vesicles to the study of Lo/Ld phase spinodal decomposition and its modulation by the ganglioside GM1. Langmuir **27**, 15074–15082 (2011)
32. Mascalchi, P., Haanappel, E., Carayon, K., Mazères, S., Salomé, L.: Probing the influence of the particle in Single Particle Tracking measurements of lipid diffusion. Soft Matter **8**, 4462 (2012)
33. Sowa, Y., Steel, B.C., Berry, R.M.: A simple backscattering microscope for fast tracking of biological molecules. Rev. Sci. Instrum. **81**, 113704 (2010)
34. Lindfors, K., Kalkbrenner, T., Stoller, P., Sandoghdar, V.: Detection and spectroscopy of gold nanoparticles using supercontinuum white light confocal microscopy. Phys. Rev. Lett. **93**, 037401 (2004)
35. Jacobsen, V., Stoller, P., Brunner, C., Vogel, V., Sandoghdar, V.: Interferometric optical detection and tracking of very small gold nanoparticles at a water-glass interface. Opt. Express **14**, 405–414 (2006)
36. Zhang, L., et al.: Interferometric detection of single gold nanoparticles calibrated against TEM size distributions. Small (2015)
37. Züchner, T., Failla, A.V., Steiner, M., Meixner, A.J.: Probing dielectric interfaces on the nanoscale with elastic scattering patterns of single gold nanorods. Opt. Express **16**, 14635–14644 (2008)
38. Ortega Arroyo, J., et al.: Label-free, all-optical detection, imaging, and tracking of a single protein. Nano Lett. **14**, 2065–2070 (2014)

39. Piliarik, M., Sandoghdar, V.: Direct optical sensing of single unlabelled proteins and super-resolution imaging of their binding sites. Nat. Commun. **5**, 4495 (2014)
40. van Dijk, M.A., Lippitz, M., Orrit, M.: Far-field optical microscopy of single metal nanoparticles. Acc. Chem. Res. **38**, 594–601 (2005)
41. van Dijk, M.A., et al.: Absorption and scattering microscopy of single metal nanoparticles. Phys. Chem. Chem. Phys. **8**, 3486 (2006)
42. Boyer, D., et al.: Photothermal imaging of nanometer-sized metal particles among scatterers. Science **297**, 1160–1163 (2002)
43. Berciaud, S., Cognet, L., Blab, G.A., Lounis, B.: Photothermal heterodyne imaging of individual nonfluorescent nanoclusters and nanocrystals. Phys. Rev. Lett. **93** (2004)
44. Ignatovich, F., Novotny, L.: Real-time and background-free detection of nanoscale particles. Phys. Rev. Lett. **96**, 013901 (2006)
45. Hong, X., et al.: Background-free detection of single 5 nm nanoparticles through interferometric cross-polarization microscopy. Nano Lett. **11**, 541–547 (2011)
46. Curtis, A.: The mechanism of adhesion of cells to glass a study by interference reflection microscopy. J. Cell. Biol. **20**, 199–215 (1964)
47. Ploem, J.S.: Reflection-contrast microscopy as a tool for investigation of the attachment of living cells to a glass surface. In: Mononuclear Phagocytes in Immunity, Infection and Pathology, pp. 405–421 (1975)
48. Monzel, C., Fenz, S.F., Merkel, R., Sengupta, K.: Probing biomembrane dynamics by dual-wavelength reflection interference contrast microscopy. Chem. Phys. Chem. **10**, 2828–2838 (2009)
49. Atilgan, E., Ovryn, B.: Reflectivity and topography of cells grown on glass-coverslips measured with phase-shifted laser feedback interference microscopy. Biomed. Opt. Express **2**, 2417–2437 (2011)
50. Matsuzaki, T. et al.: High contrast visualization of cell–hydrogel contact by advanced interferometric optical microscopy. J. Phys. Chem. Lett. 253–257 (2013)
51. Rädler, J.O., Sackmann, E.: Imaging optical thicknesses and separation distances of phospholipid vesicles at solid surfaces. Journal de Physique **II**(3), 727–748 (1993)
52. Raedler, J., Sackmann, E.: On the measurement of weak repulsive and frictional colloidal forces by reflection interference contrast microscopy. Langmuir **8**, 848–853 (1992)
53. Wiegand, G., Neumaier, K.R., Sackmann, E.: Microinterferometry: three-dimensional reconstruction of surface microtopography for thin-film and wetting studies by reflection interference contrast microscopy (RICM). Appl. Opt. **37**, 6892–6905 (1998)
54. Schilling, J., Sengupta, K., Goennenwein, S., Bausch, A., Sackmann, E.: Absolute interfacial distance measurements by dual-wavelength reflection interference contrast microscopy. Phys. Rev. E **69**, 021901 (2004)
55. Kukura, P., et al.: High-speed nanoscopic tracking of the position and orientation of a single virus. Nat. Methods **6**, 923–927 (2009)
56. Considine, P.S.: Effects of coherence on imaging systems. JOSA **56**, 1001–1007 (1966)
57. Dulin, D., Barland, S., Hachair, X., Pedaci, F.: Efficient illumination for microsecond tracking microscopy. PLoS ONE **9**, e107335 (2014)
58. Spillane, K.M., et al.: High-speed single-particle tracking of GM1 in model membranes reveals anomalous diffusion due to interleaflet coupling and molecular pinning. Nano Lett. **14**, 5390–5397 (2014)
59. Lin, Y.-H., Chang, W.-L., Hsieh, C.-L.: Shot-noise limited localization of single 20 nm gold particles with nanometer spatial precision within microseconds. Opt. Express **22**, 9159 (2014)
60. Kusumi, A., et al.: Paradigm shift of the plasma membrane concept from the two-dimensional continuum fluid to the partitioned fluid: high-speed single-molecule tracking of membrane molecules. Annu. Rev. Biophys. Biomol. Struct. **34**, 351–378 (2005)
61. Yildiz, A., Tomishige, M., Vale, R. D., Selvin, P. R.: Kinesin walks hand-over-hand. Science **303**, 676–678 (2004)
62. Sakamoto, T., Webb, M.R., Forgacs, E., White, H.D., Sellers, J.R.: Direct observation of the mechanochemical coupling in myosin Va during processive movement. Nature **455**, 128–132 (2008)

63. Kukura, P., Celebrano, M., Renn, A., Sandoghdar, V.: Imaging a single quantum dot when it is dark. Nano Lett. **9**, 926–929 (2008)
64. Ohmachi, M., et al.: Fluorescence microscopy for simultaneous observation of 3D orientation and movement and its application to quantum rod-tagged myosin V. Proc. Natl. Acad. Sci. USA **109**, 5294–5298 (2012)
65. Sönnichsen, C., Alivisatos, A.P.: Gold nanorods as novel nonbleaching plasmon-based orientation sensors for polarized single-particle microscopy. Nano Lett. **5**, 301–304 (2005)
66. Xiao, L., Qiao, Y., He, Y., Yeung, E.S.: Three dimensional orientational imaging of nanoparticles with darkfield microscopy. Anal. Chem. **82**, 5268–5274 (2010)
67. Marchuk, K., Fang, N.: Three-dimensional orientation determination of stationary anisotropic nanoparticles with sub-degree precision under total internal reflection scattering microscopy. Nano Lett. **13**, 5414–5419 (2013)
68. Enoki, S., et al.: High-speed angle-resolved imaging of a single gold nanorod with microsecond temporal resolution and one-degree angle precision. Anal. Chem. **87**, 2079–2086 (2015)
69. Züchner, T., Failla, A.V., Hartschuh, A., Meixner, A.J.: A novel approach to detect and characterize the scattering patterns of single Au nanoparticles using confocal microscopy. J. Microsc. **229**, 337–343 (2008)
70. Chojnacki, J., et al.: Maturation-dependent HIV-1 surface protein redistribution revealed by fluorescence nanoscopy. Science **338**, 524–528 (2012)
71. Otto, O., et al.: High-speed video-based tracking of optically trapped colloids. J. Opt. **13**, 044011 (2011)
72. Speidel, M., Jonas, A., Florin, E.-L.: Three-dimensional tracking of fluorescent nanoparticles with subnanometer precision by use of off-focus imaging. Opt. Lett. **28**, 69–71 (2003)
73. Toprak, E., et al.: Defocused orientation and position imaging (DOPI) of myosin V. Proc. Natl. Acad. Sci. USA **103**, 6495–6499 (2006)
74. Toprak, E., Balci, H., Blehm, B.H., Selvin, P.R.: Three-dimensional particle tracking via bifocal imaging. Nano Lett. **7**, 2043–2045 (2007)
75. Juette, M.F., Bewersdorf, J.: Three-dimensional tracking of single fluorescent particles with submillisecond temporal resolution. Nano Lett. **10**, 4657–4663 (2010)
76. Dalgarno, H.I.C., et al.: Nanometric depth resolution from multi-focal images in microscopy. J. R. Soc. Interface **8**, 942–951 (2011)
77. Holtzer, L., Meckel, T., Schmidt, T.: Nanometric three-dimensional tracking of individual quantum dots in cells. Appl. Phys. Lett. **90**, 053902 (2007)
78. Mlodzianoski, M.J., Juette, M.F., Beane, G.L., Bewersdorf, J.: Experimental characterization of 3D localization techniques for particle-tracking and super-resolution microscopy. Opt. Express **17**, 8264–8277 (2009)
79. Pavani, S.R.P., et al.: Three-dimensional, single-molecule fluorescence imaging beyond the diffraction limit by using a double-helix point spread function. Proc. Natl. Acad. Sci. USA **106**, 2995–2999 (2009)
80. Izeddin, I., et al.: PSF shaping using adaptive optics for three-dimensional single-molecule super-resolution imaging and tracking. Opt. Express **20**, 4957–4967 (2012)
81. Quirin, S., Pavani, S.R.P., Piestun, R.: Optimal 3D single-molecule localization for superresolution microscopy with aberrations and engineered point spread functions. Proc. Natl. Acad. Sci. USA **109**, 675–679 (2012)
82. Shechtman, Y., Weiss, L.E., Backer, A.S., Sahl, S.J., Moerner, W.E.: Precise three-dimensional scan-free multiple-particle tracking over large axial ranges with tetrapod point spread functions. Nano Lett. **15**, 4194–4199 (2015)
83. Mojarad, N.M., Krishnan, M.: Measuring the size and charge of single nanoscale objects in solution using an electrostatic fluidic trap. Nat. Nanotechnol. **7**, 448–452 (2012)
84. Mojarad, N., Sandoghdar, V., Krishnan, M.: Measuring three-dimensional interaction potentials using optical interference. Opt. Express **21**, 9377–9389 (2013)
85. Feng, S., Winful, H.G.: Physical origin of the Gouy phase shift. Opt. Lett. **26**, 485 (2001)
86. Hwang, J., Moerner, W.E.: Interferometry of a single nanoparticle using the Gouy phase of a focused laser beam. Opt. Commun. **280**, 487–491 (2007)

87. Selmke, M., Cichos, F.: Energy-redistribution signatures in transmission microscopy of Rayleigh and Mie particles. J. Opt. Soc. Am. A **31**, 2370–2384 (2014)
88. Bohren, C.F., Huffman, D.R.: Absorption and Scattering of Light by Small Particles, pp. 82–129 (1983)
89. Kaz, D.M., McGorty, R., Mani, M., Brenner, M.P., Manoharan, V.N.: Physical ageing of the contact line on colloidal particles at liquid interfaces. Nat. Mater. **11**, 138–142 (2011)
90. Ballard, N., Bon, S.A.F.: Equilibrium orientations of non-spherical and chemically anisotropic particles at liquid-liquid interfaces and the effect on emulsion stability. J. Colloid Interface Sci. **448**, 533–544 (2015)
91. Erickson, H.P.: Size and shape of protein molecules at the nanometer level determined by sedimentation, gel filtration, and electron microscopy. Biol. Proced. Online **11**, 32–51 (2009)
92. Ewers, H., et al.: Label-free optical detection and tracking of single virions bound to their receptors in supported membrane bilayers. Nano Lett. **7**, 2263–2266 (2007)
93. Andrecka, J., Spillane, K.M., Ortega Arroyo, J., Kukura, P.: Direct observation and control of supported lipid bilayer formation with interferometric scattering microscopy. ACS Nano **7**, 10662–10670 (2013)
94. Govorov, A.O., et al.: Gold nanoparticle ensembles as heaters and actuators: melting and collective plasmon resonances. Nanoscale Res. Lett. **1**, 84–90 (2006)
95. Zeng, N., Murphy, A.B.: Heat generation by optically and thermally interacting aggregates of gold nanoparticles under illumination. Nanotechnology **20**, 375702 (2009)
96. Qin, Z., Bischof, J.C.: Thermophysical and biological responses of gold nanoparticle laser heating. Chem. Soc. Rev. **41**, 1191 (2012)
97. Huschka, R., et al.: Light-induced release of DNA from gold nanoparticles: nanoshells and nanorods. J. Am. Chem. Soc. **133**, 12247–12255 (2011)
98. Baffou, G., Quidant, R.: Nanoplasmonics for chemistry. Chem. Soc. Rev. **43**, 3898–3907 (2014)

Chapter 3
Experimental Methods

This chapter outlines the experimental methods used throughout this thesis. All work presented in this chapter was performed by myself.

3.1 Experimental Optics and Hardware

The work presented in this chapter and all subsequent chapters were mainly performed on an optical setup that possessed: three iSCAT channels, a sample focus feedback channel and many optional fluorescence channels as described in Fig. 3.1a. This arrangement allowed iSCAT to serve as a tool to investigate nanoscopic dynamics over more than five orders of magnitude in time, under off- or on-resonant conditions for gold nanoparticles, and across multiple fields-of-view varying from the diffraction-limited to $30 \times 30 \, \mu m^2$ areas. The focus feedback channel converted the detected signal into the axial position between coverslip and sample, which was subsequently used to stabilise the focus position to within a nanometre. Lastly, the optional fluorescence channels were used for correlative experiments and to test artefacts induced by larger-sized labels.

The microscope was assembled on a table with active stabilisation, which minimised the coupling of low frequency noise from the environment to the microscope system and allowed for sub-nm tracking precision. The objective, the most sensitive optic of the setup, was fixed to a heavy stage to dampen vibrations and improve the overall stability in the measurements. This was implemented by boring a hole into an aluminium breadboard held by four large pedestals and then threading the hole with a mount complementary to the objective. An Olympus PLAPON 60X0 infinity-corrected oil-immersion objective was chosen for this setup due to its high numerical aperture (1.42) and large back-aperture associated with the relatively low $60\times$ magnification. The preference for a high numerical aperture was based on an

© Springer International Publishing AG 2018
J. Ortega Arroyo, *Investigation of Nanoscopic Dynamics and Potentials by Interferometric Scattering Microscopy*, Springer Theses,
https://doi.org/10.1007/978-3-319-77095-6_3

intrinsically higher collection efficiency of scattered light; whereas the large back-aperture facilitated the alignment process. To suppress background light and beam pointing instabilities associated with temperature gradients and air currents, the whole setup was enclosed into compartments.

As coherent illumination sources, three laser diodes each with a distinct wavelength ($\lambda = 445, 532, 635$ nm) and without a fan unit were selected on the basis of their lower coherence length, mechanical stability and cost-effectiveness in terms of power output relative to other laser modules. The blue laser excitation was primarily used for label-free and wide-field-of-view experiments, due to the increased scattering amplitude from non-resonant objects as expected from a $1/\lambda^2$ dependence. The green and a portion of the red excitations were mostly applied for high-speed on- and off-resonant imaging of gold nano-particles. The remaining fraction of the red excitation served as the feedback channel input.

As detectors in the interferometric channels, two high-speed CMOS cameras (Photon Focus MV-D1024-160-CL-8, Lachen, Switzerland) were chosen because of their higher achievable frame rates, versatility in the selection of region of interests, large full-well capacities, and generally lower cost than their EM-CCD counterparts. Specifically, the main selection criteria were the minimum frame time of 19 μs and the large full-well capacity, i.e. the number of photo-electrons each pixel can detect before saturation. These criteria defined the choice of magnification, and, as a result, the maximum achievable sensitivity per image.

For most applications, the aim was to achieve pixel-to-pixel signal variations on the order of 0.1% with an effective pixel size on the order of 80–100 nm. Under the assumption of a shot noise-limited measurement, 0.1% variations imply the collection of 10^6 photons per pixel. Unfortunately, most CMOS sensors lack such dynamic range and instead possess full-well capacities in the tens to hundreds of thousands of photoelectrons. Nevertheless, pixel binning can increase the full-well capacity to several millions of photoelectrons by making an effective super-pixel. For the Photon Focus camera the well depth of 200,000 photoelectrons was turned into one approaching 10^6 photoelectrons by spatially binning 3×3 pixels. This required the magnification to be modified, such that the effective pixel size of the sensor was on the order of $100/3 = 33$ nm.

Experimentally the magnification was modified by using an imaging lens with a different focal length to the one specified by the objective as the tube lens length. In the case of the PLAPON objective and a 1000 mm imaging lens, an overall enhancement of 5.55 (1000 mm/180 mm) was achieved, thus yielding a 333× magnification, which for the pixel size of the camera of 10.6 μm translates into an effective image plane pixel dimensions of 31.8 nm. In the case of high-speed imaging applications, a magnification of 100× was sufficient. This allowed larger fields-of-view and also effectively decreased the camera frame time given that fewer pixels had to be read out.

With regards to the sample mount and translation stage, the following criteria were considered for the respective hardware components: firstly, the axial positioning of the sample should be fine-tunable and stabilised to within a couple of nanometres by active feedback; secondly, the stages should not introduce or should allow correction

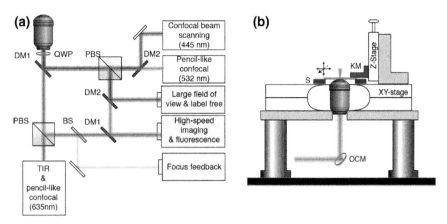

Fig. 3.1 Microscope setup. a Schematic of the setup. QWP: quarter wave-plate PBS (Thorlabs, AQWP05M-600): polarising beam-splitter (Thorlabs, PBS201), BS:beam-sampler (Thorlabs, BSF10-A), DM1: dichroic mirror (Thorlabs, DMLP605), DM2: dichroic mirror (Thorlabs, DMLP505) **b** Side-view illustration of the sample stage and objective mount. KM: kinematic mirror mount (Liop-tec, SR100-100R-2-BU), S: sample stage; OCM: objective coupling mirror (Thorlabs, BB05-E02)

for sample tilt; and thirdly, sample displacements in the image plane should not affect the focus. Thus, for lateral control, a two-axis manual translation stage (OptoSigma 122-0245, 120 mm × 120 mm) fit with 17.4 μm travel range actuators (Thorlabs, AE0505D16F) was mounted directly on the breadboard where the objective was fixed (Fig. 3.1b). For focus control, a single-axis translation stage with a differential manual adjustment and a piezo element allowing 20 μm travel-range (Thorlabs NFL5DP20) was fixed on top of the 2D translation stage with an L-bracket. As the sample mount, a 7 mm thick aluminium block with a 40 mm hole and two holding pins was used. To enable precise tilt correction, the sample mount was attached to the vertical translation stage via a high stability kinematic mirror mount.

3.1.1 Interferometric Scattering Channel

Implementation of the iSCAT channels is composed of three main phases: alignment of the excitation arm in the absence of any lenses other than the objective, alignment of the detection arm, and placement of the remaining lenses. For the sake of generality I first outline the overall process for an non-scanned illumination achieved by underfilling the objective with a collimated beam; and then elaborate on the specific details involved with Koehler and confocal beam scanning illumination schemes.

Regardless of whether the illumination is scanned or non-scanned, the laser output was cleaned and turned into a Gaussian-like mode by either coupling into a single-mode fibre or spatially filtering the beam through a pinhole. This served the role of transforming all beam instabilities into intensity fluctuations, which can be accounted for by normalisation. The cleaned beams were then collimated to a diameter of approximately 4–5 mm full-width at half-maximum. The small diameter ensured that the beam under-filled the objective and was smaller than the acoustic aperture size of the acousto-optical beam deflectors (5 mm). Special care was taken to have all beams travel straight and in a plane parallel to the optical table before introducing the objective coupling mirror, i.e. a mirror placed at a 45° angle.

As the critical step, the beam reflected by the objective coupling mirror was aligned so that it travelled straight and through the centre of the objective. To achieve a straight vertical beam-path a pendulum was hung from the ceiling directly above the objective coupling mirror and a target was marked in its place (ceiling). The ceiling target served as a rough guide to direct the beam vertically upwards with the objective coupling mirror. To centre the objective onto the beam, firstly a long focal length lens was introduced into the excitation arm without causing beam displacements. Next, the objective was inserted into the vertical beam-path and moved until no beam displacements were observed relative to the ceiling target. If the beam projected onto the ceiling was too large, the position of the long focal length lens was changed until the smallest area was projected onto the ceiling. At this point it was deemed that the lens focused the beam into the back focal plane (BFP) of the objective.

As a fine-tuning step in the absence of the BFP lens, the objective was removed and a flat mirror was inserted in its place with the premise that if the vertical beam path was straight, the reflection from the mirror would overlap with the excitation arm (the beam would come back on itself). If not, only the objective coupling mirror was adjusted until both beams overlapped. Then, with the vertically-placed mirror removed, the target in the ceiling was updated to the position projected by the beam. Finally, the objective was re-centred by iterating through the previously mentioned procedure. Alignment of the microscope system, to be interpreted as the incident beam travelling straight through the centre of the objective, was checked by imaging a coverslip sample and looking at the resulting beam profile in the detection arm with a camera. The microscope was deemed correctly aligned when the profile of the beam focused concentrically, with radial symmetry and without lateral displacements as the sample was brought in-and out-of-focus.

The detection channel was aligned by first picking off the reflection from the beamsplitter, in this case the image, and ensuring that the beam travelled straight and in the plane parallel to the table with the aid of at least two mirrors mounted on high precision kinematic mounts. Next, the cameras were aligned such that the beam hit precisely the middle of the sensor chip, and the plane of the sensor was perpendicular to the imaging path. With the choice of magnification in mind, the imaging lenses were introduced at the appropriate focal distance away from the camera sensor. It is worth remarking that the imaging lenses were placed as close as possible to the objective to improve the scattering light collection and reduce the loss of numerical aperture. Proper positioning of the lenses was met by two criteria of the beam profile imaged on the camera. On one hand, the distance between the lens and the camera sensor was deemed correct when the beam profile had the largest intensity and smallest area for an incident collimated beam, a process known

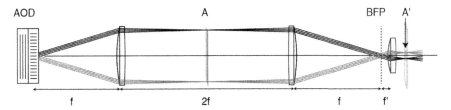

Fig. 3.2 Telecentric imaging. Ray tracing diagram of the 4f-geometry involved in a telecentric relay imaging system. A: area scanned with the AODs; A': scanned area in the image plane

as infinity-correction. On the other hand, correct placement of the lens to minimise introduction of optical aberrations, such as coma and astigmatism, was verified when the centre of mass of the focused beam did not change relative to beam profile in the absence of the imaging lens.

Finally with the detection and excitation arms aligned, the remaining lenses were introduced one-by-one into the beam path always verifying that no beam displacements had occurred. For example, a single lens in the excitation arm, positioned a focal distance away from the BFP of the objective switched the non-scanned illumination from pencil-like confocal to Koehler. For larger fields-of-view in Koehler illumination, the initial beam diameter was expanded by a telescope composed of a lens-pair.

In the case of beam scanning confocal illumination, three additional optical elements were introduced: a telecentric lens system, two acousto-optical beam deflectors (Gooch and Housego, 45100-5-6.5Deg-.51-XY) and two mirrors fixed into high-precision kinematic mounts. The telecentric lens system was composed of two 400 mm focal length lenses arranged in a 4-f geometry with respect to the BFP of the objective and the middle point between the two AODs as depicted in Fig. 3.2. This lens system relayed the beam deflections from the AODs into the BFP of the objective.

The AOD alignment was achieved in three steps. First, the collimated output from the laser was directed in a straight line and in a plane parallel to the optical table. Second, one AOD was placed in the beam path and then the angle of the crystal element was adjusted so that the output of the first diffraction order was maximised to about 70% of the incident power. Note that the AODs only efficiently diffract if the incident polarisation is parallel to the acoustic wave propagation, and the output polarisation is rotated by 90°, so introduction of a half-wave plate before the first AOD may be required to achieve the optimal transmission efficiency. Third, the remaining AOD was positioned and its output power optimised by adjusting the angle of incidence between the face of the crystal element and the first order diffraction from the first AOD. The two adjustable mirrors were placed immediately after the AODs to compensate for the change in height and direction associated with the AODs, and thereby ensure that the non-scanned deflected beam travels straight and in a plane parallel to the table.

3.1.2   Focus Control Feedback Channel

Two different versions of the feedback channel for focus stabilisation were implemented based on the principle that changes in focus position, the distance between
the sample and the objective, translate into displacements of the profile of a total
internally reflected beam imaged onto a camera (Fig. 3.3). In the first version, a
fraction of the collimated output from a single-mode fibre-coupled red laser was
picked off using a beam-sampler and then focused onto the BFP of the objective
using a combination of a lens, another beam-sampler, and two adjustable mirrors,
one of which was mounted on a linear translation stage. Total-internal-reflection in
a slightly off-axis orientation was achieved by translation of the stage, followed by a
small re-adjustment in the deflection angle of the mirror mounted on the stage. The
total-internal-reflection was then imaged onto a camera (PointGrey, Firefly USB 2
CMOS) with a cylindrical lens rather than a normal lens to increase the SNR in determining the centre position of the beam. It was crucial that the TIR was achieved in an
off-axis position, otherwise the cylindrical lens would make the projected image on
the camera insensitive to lateral beam displacements. Information regarding the focus
position was encoded to a specific beam displacement on the camera (Fig. 3.3b). This
position was determined by calculating the centre of mass of the linear profile of the
beam after summing over all camera rows. As a limitation, this feedback channel
was very sensitive to beam-pointing instabilities and air currents, as these caused
unwanted beam displacements, which could be mistaken for changes in focus.

In the second and simpler version, a collimated beam, with a size comparable to
the back-aperture of the objective, was directed into the objective. The resulting back-

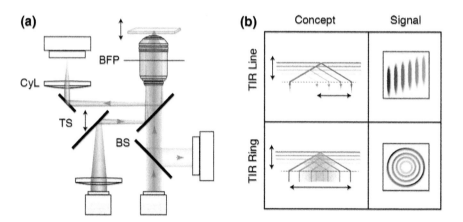

Fig. 3.3 Focus feedback channel. a Schematic of the two different versions of feedback channels
implemented in this setup. Cyl: Cylindrical lens, TS: Translation stage, BS: beam sampler. **b** Concept
and expected signal detected in each feedback channel. Ray tracing diagrams indicate total-internal-
reflection at different positions of the coverslip. Vertical arrows: axial displacement; Horizontal
arrows: corresponding axial displacements detected in the feedback channel

reflection signal was dominated in intensity by the total-internal-reflection, resulting in a ring-like profile. Proper alignment of the incident beam corresponded to when the ring profile had a radially symmetric intensity profile. This ring profile was then imaged onto the camera. In this case, the focus position was encoded within the diameter of the ring; thus eliminating beam-pointing artefacts (Fig. 3.3b). The ring diameter was determined by first finding the centre of mass of the ring via a radial symmetry algorithm, then mapping all the pixel positions to polar coordinates, and finally finding the centre of mass of the radial coordinate distribution, which results in a similar linear profile to the previous version.

3.1.3 Single-Molecule Fluorescence Channel

One of the great advantages of iSCAT is its compatibility with fluorescence and its ease of implementation on already existing custom-built or commercially available epi-fluorescence or TIRF microscopes. In principle, a single dichroic mirror in the beam path can be used to couple in a fluorescence channel, as long as it can separate the wavelength used for iSCAT from both fluorescence excitation and emission spectra. In the current setup, multiple implementations of fluorescence microscopy were possible, given the range of dichroics and types of illumination: epi-, total-internally-reflected and confocal beam scanning fluorescence. The fluorescence emission was detected by the addition of a bandpass filter in the high-speed imaging channel. To achieve total-internal-reflection illumination at the glass-water interface, the excitation source was focused into the back focal plane of the microscope objective and displaced laterally.

3.1.4 Sample Stage Stabilisation

For axial stabilisation, first a calibration between applied voltage to the piezo element in the stage and the feedback signal was performed by stepping through the focus in discrete amounts and allowing the collection of at least 20 data-points per position. The applied voltage was linearly mapped to a corresponding focus position. The average feedback value recorded for each step was plotted against the mapped focus position and fit to a linear model. This calibration then, enabled the relation of the feedback signal to a focus position in real time. For focus stabilisation, a user-selected position was locked onto by initially setting: a target position value, z_0; a minimum threshold criteria for sensitivity, δz; a gain parameter, g; and a linear transformation function that mapped the change in position into an input voltage, $h(z)$. Here the gain parameter determines how quickly the feedback loop converges to the target value, and is analogous to tuning the damping ratio parameter for a system modelled as a damped harmonic oscillator [1]. If the gain value was too high, oscillatory behaviour was observed typical of an underdamped system; if it was too low, the response time was too slow to actively stabilise the focus, analogous to an overdamped system. This parameter was determined experimentally by trial and error and was dependent on the active feedback rate. The active feedback function for applied voltage to the piezo element can thus be described by the following pseudo code:

$$f(V) = \begin{cases} h(g \times (z - z_0)) : |z - z_0| \geq \delta z \\ 0 \qquad\qquad\qquad : |z - z_0| < \delta z \end{cases}$$

The same procedure was followed for lateral stabilisation, with the exception that the sample was stepped within the image plane rather than through the focus. The feedback signal was either extracted from the position of one or several immobilised objects in the sample or via the cross-correlation between a target and a probe image, a topic further detailed in the self-referencing subsection.

3.1.5 Camera Characterisation

The response of the camera to light can be summarised by three major parameters: the gain (k), the well depth and the total noise. Together all these three parameters determine the sensitivity and overall performance of an experiment. These parameters were extracted by calculating the photon transfer curve of the camera [2]. The photon transfer curve is based on the analysis of the total noise of the image as a function of different photon fluxes, under the assumption that the total noise of an image is composed of three major sources: read-out, related to the electronics of the camera; shot, related to stochastic nature of light; and fixed pattern, related to temporally invariant fluctuations in the photon response of each pixel. All these sources add together in quadrature to provide:

$$\sigma_{tot}^2 = \sigma_R^2 + \sigma_S^2 + \sigma_{FPN}^2 \tag{3.1}$$

Given that a camera records the signal intensity, S, in arbitrary digital units, the fixed pattern noise is quantified as a fraction, P_N, of the signal, and the total photon count per pixel is related to the gain by $N = kS$. The above relation can be expressed in digital units as a function of signal intensity:

$$\sigma_{tot}^2(S) = \sigma_R^2 + \frac{1}{k}S + P_N^2 S^2 \tag{3.2}$$

Moreover if two images are taken under the same illumination conditions, the fixed pattern noise is constant and the total noise in the resulting differential image is:

$$\sigma_{dif}^2(S) = 2\left(\sigma_R^2 + \frac{1}{k}S\right) \tag{3.3}$$

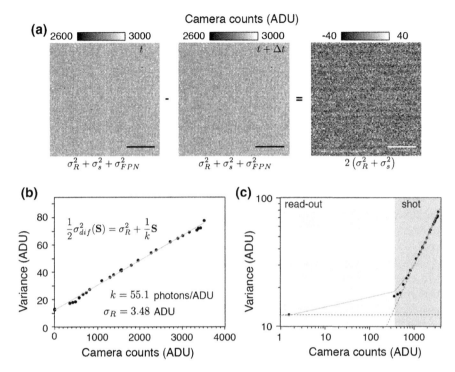

Fig. 3.4 Characterisation of the photon response of the camera. a Noise contributions present in the difference between two evenly illuminated images offset in time by Δt. Scale bars: 2 μm. **b** Photon transfer curve obtained by calculating the variance in signal from differential images taken at different illumination intensities. Solid line represents the best fit to a straight line. **c** Same curve as in (**b**) but shown in log-log form to highlight the read-out and shot-noise dominant regions of the camera. Dotted lines refer to the pure read-out noise and shot noise contributions

From the above relation, the read-out noise, the gain of the camera, and the well depth can be extracted by fitting a linear function to a plot of the variance in the signal of a differential image as a function of different incident photon fluxes. In practice, the data for the photon transfer curve was acquired by homogeneously illuminating the camera chip with a torch and taking sequences of images at different intensities, all whilst keeping the exposure time constant. To calculate the photon transfer curve, first the camera offset counts were removed from each image. Next, pairs of images with the same average intensity value were subtracted from one another (Fig. 3.4a). Then, the variance of the differential image was determined, and the average value for pairs of images with the same intensity was calculated and plotted (Fig. 3.4b). Finally, after extracting the parameters from the photon transfer curve analysis, the shot noise-limited regime of the camera was determined by transforming the previous plot, together with the corresponding fit, into a log-log scale graph, and identifying a kink in the best-fit photon transfer curve (Fig. 3.4c).

The parameters extracted from the photon transfer curve of one of the two CMOS cameras used in all experiments were: $\sigma_R = 191$ photo-electrons/pixel, $k = 55$ photons/digital unit and a well depth of 202,000 photo-electrons, in agreement with the camera manufacturers specifications.

3.1.6 Operation and Synchronisation of the Acousto-Optic Beam Deflector

Confocal beam scanning illumination can be achieved by different types of beam deflection devices: resonant scanning mirrors, electro-optic deflectors and acousto-optic deflectors. In the setup, the acousto-optic deflectors were chosen due their higher scanning rates compared to resonant scanning mirrors and lower cost compared to electro-acoustic modulators. The scanning rates of the AODs, on the order of a couple of hundred kHz, allowed imaging with exposure times as low as 0.4 ms.

To ensure that the pattern scanned by the beam was the same for each image, which translated into a constant illumination profile, the function generators were phase-locked and operated in burst mode during the exposure time of the camera. Burst mode denotes that the beam deflections occur only for a number of cycles determined by the exposure time of the camera. To synchronise the AODs with the camera, the function generators were externally triggered by the frame-grabber of the camera each time the camera started an exposure (Fig. 3.5a). Failure to synchronise the data acquisition resulted in time-varying beat patterns in the images caused by the drift in phase between the two orthogonal beam deflection units operating at different frequencies. As an alternative, the function generators could instead trigger the camera and the AODs, thus synchronising the data acquisition process.

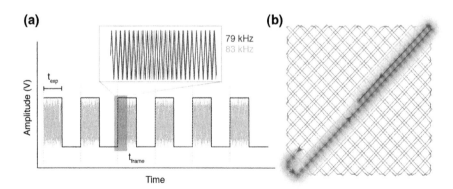

Fig. 3.5 Concept of AOD synchronisation and beam scanning illumination. a Triggering sequence for AOD synchronisation with respect to the camera frame time. The AODs only scan while the camera is exposing, and do so at slightly different frequencies. **b** Resulting scan pattern across the sample during a single exposure

Each AOD required a radio frequency module, which itself was connected to an arbitrary function generator. These function generators controlled the amplitude, offset, shape, and frequency of the scanning. Specifically for most experiments the two AOD channels were scanned in a sawtooth fashion at 83 and 79 kHz, respectively. Both the absolute and relative frequencies were chosen to induce the smallest detectable fluctuations in the background light intensity on the time-scale of the camera exposure time. For a non-scanned beam with FWHM of 1 μm few tens of scans over an area of 10×10 μm^2 were sufficient to generate a highly uniform illumination (Fig. 3.5b). At the given scan speeds, this process only takes $\sim$100 μs, much faster than the shortest exposure time. Any spot broadening induced by the limited speed of the acoustic wave in the deflector only served to further smooth the illumination.

3.1.7 Data Acquisition

Data from each camera was acquired continuously under a ring buffer mode using custom-written software in LabVIEW. The ring buffer acquisition is analogous to transferring frames into a list with a finite length, which, when filled, gets overwritten element by element (Fig. 3.6). This has the advantage that individual frames are temporarily locked and thus can be transferred to memory before being overwritten by the next frame, thus allowing for more stable data transfers and real-time image processing.

However ring buffer mode acquisition alone does not guarantee full synchronisation between the frame-grabber hardware and the camera at all frame rates, as this depends on the deterministic time-scale of the interfacing hardware. For computers this time-scale is on the order of milliseconds and for field-programmable gate arrays (FPGA) platforms, it is on the order of tens of microseconds. This can be explained

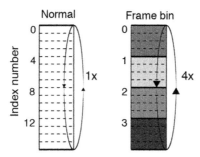

Fig. 3.6 Basics of ring-buffer data acquisition. Schematic illustrating acquisition of data via a ring buffer of 16 elements. In a normal execution, each new frame is mapped to a position within the ring. Once the whole list is full, the next frame will overwrite the first element in the ring buffer and so forth. In the case of frame binning, frames are written in blocks of size determined by the frame bin, which for the above diagram is four. The advantage of frame binning is that the acquisition and subsequent reading of the data is sped up

by considering the basics of ring buffer acquisition: to extract a frame, the computer must poll the ring buffer and find the element containing the image. However the polling rate is limited by the deterministic time-scale, and as a result the computer lags behind the camera; thus causing loss of information and synchronisation.

To circumvent this problem and to continuously acquire data at almost two orders of magnitude faster speeds than the deterministic time-scale of a computer, a hardware and a software solution were proposed based on the concept of reading multiple frames at a time. The hardware solution consisted of directly transferring data from the camera in packets containing a set amount of frames, termed frame bins. In the software solution, data was extracted from the ring buffer in frame bins by polling that data in integer multiples of the frame bin. As a result the ring buffer length was set to an integer multiple of the frame bins.

To save a sequence of images, first a fixed amount of space in the random-access memory, RAM, was previously preallocated, typically on the order of 2 GBs. Next, during the acquisition process, frames were streamed into the preallocated space in RAM before finally being saved to the hard-drive. RAM streaming was necessary because direct data transfer to the hard-drive was too slow relative to the rate of data acquisition and as a result caused loss of synchronisation between the camera and the computer.

The synchronisation between the camera and the frame grabber was benchmarked by recording a sequence of images where a beam was scanned in one direction with an AOD at two known frequencies, thus producing a beat pattern. Specifically, one of the AODs was scanned sinusoidally at a frequency of 2000 Hz with a 70% sinusoidal amplitude modulation at a frequency of 200 Hz. The beating pattern time-trace was retrieved from the sequence of acquired images from the average intensity within a region of interest (Fig. 3.7a). The input frequencies were then determined by fast Fourier transforms of the retrieved beat pattern trace (Fig. 3.7b). No frame skipping was evident when the input and retrieved frequencies matched and the higher harmonics were identified. Anomalous data acquisition was identified when the Fourier transforms contained additional peaks that did not coincide with the input frequencies or their higher harmonics. As a further proof, an intensity map obtained by dividing the time traces into evenly spaced segments was plotted (Fig. 3.7c). Frame skipping was identified by sudden phase jumps or irregularities in the beat pattern time trace.

3.2 Experimental iSCAT Microscopy

3.2.1 Image Processing

Regardless of the wide-field detection approach used in iSCAT, the acquired raw images contain static features that are intrinsic to the microscope system and yet extrinsic to the sample; thereby compromising the sensitivity of the instrument.

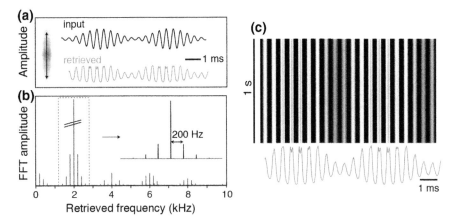

Fig. 3.7 Verification for continuous high-speed data acquisition. a Input and retrieved beating pattern from a beam scanned sinusoidally in one dimension at a frequency of 2000 Hz with a 70% sinusoidal amplitude modulation at a frequency of 200 Hz. **b** Fast Fourier transform of the retrieved time trace showing the input frequencies of 2000 and 200 Hz and their respective harmonics. **c** Intensity map obtained by dividing the beating time trace into evenly spaced 1 ms segments. Notice the continuity of the beating pattern throughout the 1 s acquisition

These features result from non-uniform illumination, non-uniform pixel response (fixed pattern noise), spurious back-reflections, unwanted interference between back-reflections and imperfections found on the optical elements of the setup. These time-invariant features do not change as a function of lateral sample displacement. Isolation and subsequent removal of these signatures was performed by a process known as flat fielding, which was performed by normalising the raw images by a temporal median image (Fig. 3.8a). The temporal median image was determined by recording a sequence of images that satisfied two specific conditions. Firstly, the focus position of the sample must remain constant as some of these signatures are focus dependant. Secondly, the speed of the lateral displacement must be sufficient so that on average each pixel does not contain information from a specific sample feature (e.g. 10 $\mu m\,s^{-1}$ at 100 Hz). Thus, computing the median for each pixel generates a representative background image lacking any sample specific features (Fig. 3.8b). Given that this image was generated from the recording of several tens to a hundred individual images, shot noise induced background fluctuations are significantly reduced compared to single acquired images. The noise levels were further minimised by binning multiple frames together before calculating the temporal median. Finally, normalisation by the temporal median image not only dramatically improves the image contrast by removing all stationary background features, but also does not introduce any additional noise into the image (Fig. 3.8c).

Flat-fielding only takes care of the static features in the background. In addition, the large dependence on focus position stability during its generation, and its application to a very narrow range of sample focus position, severely limits the extent

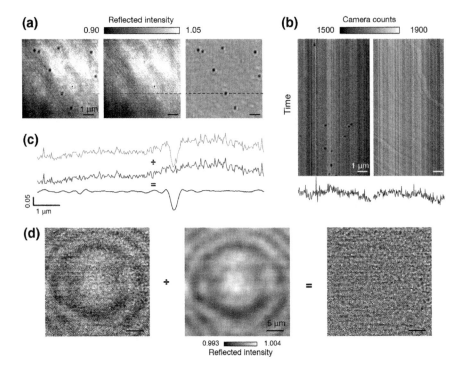

Fig. 3.8 **Flat-field correction. a** Raw, median and median-divided iSCAT images. The raw iSCAT image displays an inhomogeneous baseline due to uneven illumination and other background sources. **b** Kymographs of x- and y-cuts through the image recorded as the sample is translated left to right. Both display stripes, which are indicative of sample-independent background contributions. Scale bars: 1 μm. **c** Division of the raw (orange) by the median (blue) drastically reduces the image background as illustrated for a horizontal cut marked in (**a**). **d** Removal of additional large-scale features due to slowly varying fluctuations by pseudo flat-fielding. Scale bars: 5 μm

to which it can recover the quality of the image. Under real experimental conditions, other time-varying signal fluctuations exist, such as unsynchronised AODs, laser mode hops, sample focus drift and large out-of-focus scattering signals, that lead to inhomogeneous baselines and image corruption. In general all these features correspond to slowly varying signals extending over areas much larger than a typical point-spread function, and thus can be extracted by an additional image processing tool, which we denote pseudo flat-field (Fig. 3.8d). The pseudo flat-field uses a band pass filter obtained by setting a kernel with a size significantly larger than the typical point-spread function, and with the operation to calculate either the median or mean. The result, is an image that collects features that are larger than those defined by diffraction limited spots. Much like the flat-field case, division of the pseudo flat-field image improves the image quality and produces an image with the most homogeneous baseline.

Regardless of the above image processing, intrinsic sample features such as angstrom-level surface roughness contribute to a scattering background that limits the achievable signal to noise ratio required either for high localisation precision or bio-sensing measurements. Fortunately, this scattering background is constant and can be removed by two distinct yet complementary approaches: dynamic imaging and differential imaging (Fig. 3.9). In dynamic imaging, particles of interest that are mobile are treated as foreground elements and extracted from the rest of the image, considered as the background element. There are a myriad of algorithms that can be borrowed from the field of machine vision to differentiate background from foreground elements in an image [3–5]. In the simplest and least computationally demanding scenario, which, is when nano-objects move sufficiently enough, so that on average they spend less than half of the time on the same pixels, the static background elements can be extracted and removed by a temporal median approach. This temporal median image contains all the static features of the sample. Subtraction of this median image from the rest of the data set yields differential images dominated by shot noise and composed of dynamic features in the sample (Fig. 3.9a). For extremely weak scatterers, shot noise dominates the signal, but the SNR can be improved by pixel binning and temporal averaging as discussed above.

The effectiveness of dynamic imaging is closely related to the speed at which the object moves and the time resolution employed. To illustrate this, consider the 2D Brownian diffusion of a small label imaged at a frame rate of 1000 Hz for an observation time of 1 s. In order for temporal median filtering applied to dynamic imaging to efficiently identify and subsequently remove all static features, the mobile object must move a distance at least three times the radius of a diffraction limited spot (DLS) away from its starting point in 500 ms. For high NA objectives the full width at half maximum of a DLS is on the order of 200 nm, therefore we require the PSF of the object to have moved $(0.6 \ \mu m)^2 = 0.36 \ \mu m^2$ in 500 ms. Using the relation MSD = 4Dt we can conclude that the diffusion coefficient of the object must be on the order of $0.18 \ \mu m^2 \, s^{-1}$ or above for dynamic imaging to perform optimally under temporal median filtering.

Dynamic imaging is prone to fail when the signals of interest are immobile, for example when detecting very small particles such as single proteins in a bio-sensing assay or phase transitions at the nanoscale. If the immobile features possess the characteristic akin to an active and inactive state, such as the binding or unbinding of an analyte, differential imaging can be applied to generate shot noise-limited images with the signal of interest. Figure 3.9b shows two iSCAT images, one recorded at time t and the other at a fixed time interval, Δt, later. Subtraction of the former image from the latter removes all stationary features from the image and reveals any dynamics in the sample occurring during the specified interval. Sample drift and time-varying signals from background scatterers limit both approaches, but as long as these processes happen in a time-scale much slower than the rate of generation of the scattering background they are considered to be inconsequential.

Ultimately, image processing does the critical task of transforming the raw data into shot noise-limited images containing diffraction limited spots that greatly resemble data acquired under a state-of-the-art single-molecule fluorescence microscope.

(a)

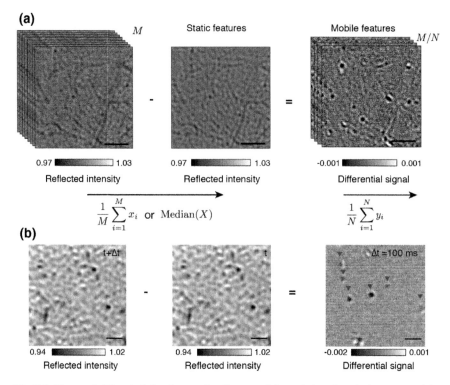

$$\frac{1}{M}\sum_{i=1}^{M} x_i \text{ or Median}(X)$$ $$\frac{1}{N}\sum_{i=1}^{N} y_i$$

Fig. 3.9 Removal of the static background. a Concept of dynamic imaging: An image containing purely stationary iSCAT features, obtained by taking the median or averaging over a sequence containing M iSCAT images, is subtracted from the raw image sequence to yield a set of images containing only mobile features. The resultant images with mobile features can then be averaged in N bins, to increase the SNR. Note the 30-fold reduction in the intensity scale of the final image. Sample type: actin filaments immobilised on a glass surface with processive myosin 5a molecules in the presence of ATP. **b** Concept of differential imaging: sets of images offset by a time Δt are subtracted to yield differential images that contain features that only appeared within the said time interval. These differential images are also averaged to increase the SNR. Sample type: deposition of α-synuclein monomers onto a bare glass coverslip. Scale bars: 1 μm

As a result, decades of research and development of algorithms for spot detection, super-resolution and trajectory linking can be directly transferred and applied to iSCAT [6, 7].

3.2.2 Spot Detection

From the processed images, diffraction limited spots were identified by a sequence of pattern recognition operations that output the location of a candidate feature to the nearest integer pixel value. First, each image was filtered using a Gaussian kernel with a size defined by the FWHM of the diffraction limited spots ($\sim$200 nm). This Gaussian blurring step artificially enhances the SNR of such features by smoothing out high frequency noise represented by shot noise. Next, a simple cluster analysis known as the non-maximum suppression algorithm was implemented to only identify the local extrema within the optical resolution of the imaging system [8]. This algorithm consists of dividing the image into $(n+1) \times (n+1)$ blocks, with n corresponding to the diffraction limit size in pixels, and finding the maximum value and its corresponding pixel position, (m_i, m_j), within each one. The value from the pixel position (m_i, m_j) is defined as a local maximum when no higher value within the neighbourhood of $[m_i - n, m_i + n] \times [m_j - n, m_j + n]$ is found.

To remove local maxima attributed to noise, only pixel values exceeding twice the standard deviation of the entire image in the absence of outliers were selected. The standard deviation of the image was estimated by the median absolute deviation due to its robustness against outliers using by the following expression:

$$\sigma \approx 1.4826 \, \text{med}_i |x_i - \text{med}_j x_j| \tag{3.4}$$

where, $x_{i,j}$, represent the values of each pixel and $\text{med}_{i,j}$, the median taken over the respective index. The remaining pixel coordinates satisfying the median absolute deviation criteria were further filtered by a range of user-defined contrasts. Finally candidate particles were segmented into regions of interest corresponding to approximately 1 μm^2, which for a magnification of 100$\times$ corresponded to 9×9 pixels (effective pixel size of 106 nm).

3.2.3 Localisation

Sub-pixel localisation information was obtained from the candidate features, present in each segmented region of interest, by a non-linear least square fit to an elliptical 2D Gaussian with a constant offset expressed by the following functional form:

$$G(\mathbf{x}, \mathbf{y}) = C \, e^{-\frac{1}{2}\left(\frac{x-x_0}{s_x}\right)^2} e^{-\frac{1}{2}\left(\frac{y-y_0}{s_y}\right)^2} + B. \tag{3.5}$$

with C representing the signal contrast, x_0 and y_0 the centre of mass location in each axis, s_x and s_y the width in each axis and B a constant baseline. The equation is written in such a way to indicate the separability of the functional form, which was exploited to optimise the execution speed of the process from $O(MN)$ to $O(M + N)$, where M and N refer to the size of the image in each dimension [9]. The non-linear least square fit was performed by the iterative Levenberg-Marquardt algorithm, which requires a "good" initial guess for each of the floating parameters.

The initial guess was determined by a faster localisation algorithm developed by Parthasarathy, termed radial symmetry centres [10]. This approach exploits the property that the centre of a radially symmetric intensity distribution, in this case the PSF, can be determined geometrically as the intersection of the lines parallel to the calculated gradient at each point of the image. The advantage of this approach over iterative ones is that it is model-free and requires a single iteration; which leads to an improvement in the execution time by two orders of magnitude, all whilst achieving similar performance to model-based techniques. Poor localisations were rejected on the basis of their FWHM and eccentricity. Here the eccentricity is defined as: $\sqrt{1 - (s_1/s_2)^2}$ such that $s_2 > s_1$, and a value of 0 and 1 correspond to a circle and parabola respectively.

3.2.4 Trajectory Linking

The process of finding the temporal correspondence between the localisations of each image with a sequence of frames is known as trajectory linking and is an integral part of SPT. Of all the algorithms, the multiple-hypothesis tracking, is the most accurate owing to the computation of all possible paths given a set of particles and a number of frames. Despite its global optimisation in space and time, multiple hypothesis tracking is computationally prohibitive for systems with more than a few tens of particles and hundreds of time-points. Given that most iSCAT measurements far exceed these values, heuristic approaches (based on experience), were implemented due to their much lower computational cost. The simplest trajectory linking algorithm used in the analysis of single-molecule data from iSCAT measurements was classified as greedy, where only the local, rather than the global connectivity between adjacent frames is optimised.

The local optimum was found by minimising the distance between particles from one frame to another, rather than throughout the whole sequence of acquired frames. This was achieved by calculating a distance cost matrix between consecutive frames and selecting the minimum across each row as the connectivity assignment. When the minimum distance for any row exceeded a search radius value, chosen to reflect the dynamics of the system under study, a no linking option was preferred. Given these conditions three outcomes were possible: track initiation, track linking and track termination. Track initiation occurred whenever features in frame $t + 1$ had no links to one in frame t or for $t = 0$. Track linking followed when a feature in $t + 1$ linked to one in frame t. Finally, track termination occurred when no features in present in frame $t + 1$ linked back to frame t.

Gaps in the trajectories were closed by merging track segments that were within a search radius given the temporal separation between them [11]. Tracks with poor connectivity were identified due to their short length and rejected from the analysis. One of the main limitations of the greedy approach is that a one-to-one mapping between features of consecutive frames is not guaranteed. This can be resolved, however, by either solving the linear assignment problem for the cost matrix [12–14], which consists of finding the minimum cost assignment of features from one frame to another such that each assignment is unique, or eliminating repeated tracks on the basis of similarity.

3.2.5 Assessment of Localisation Precision

To assess the shot noise-limited behaviour of the setup, two adjacent immobilised particles were localised and their positions plotted as a function of time (Fig. 3.10a). Drift contributions were removed by comparing the fluctuations in the distance between the two particles (Fig. 3.10b). The nominal localisation precision was determined from the spread in the distribution of the distance fluctuations (Fig. 3.10c, inset). By measuring the localisation precision at different illumination intensities (or exposure times) the shot noise-limited dependence was evaluated and served as a reference for future measurements (Fig. 3.10c).

3.2.6 Self-referencing

Extrinsic noise in the sample area constitutes a prevalent problem for single-molecule techniques and iSCAT is not the exception, as this type of noise makes an experiment depart from shot noise, and introduces artefacts in the analysis of single particle tracks. The most common sources for this type of noise correspond to thermal/mechanical drift and external vibrations being coupled into the sample area. Although this noise can be significantly reduced by simply using more expensive hardware; for instance in the form of better vibration isolation optical tables, smoother translation stages with strain gauges, and acoustically and thermally isolating the

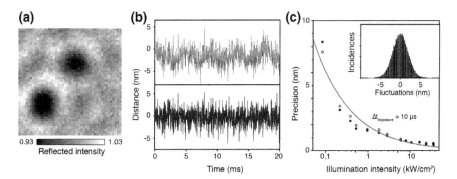

Fig. 3.10 Determination of the nominal localisation precision. a Flat-field corrected iSCAT image of 20 nm gold nanoparticles immobilised on a glass coverslip. Small variations in the background signal that move together with sample displacements correspond to imperfections of the substrate. **b** The top trace shows the fitted position of both gold nanoparticles as a function of time. The modulation in both traces is caused by vibrations extrinsic to the sample. The bottom trace shows the fluctuation in the distance between the two nanoparticles as a function of time. **c** Localisation precision as a function of illumination intensity for 40 nm gold nanoparticles. An increase in illumination intensity allows higher localisation precision for a given exposure time. The solid line shows the expected behaviour for a shot noise-limited process

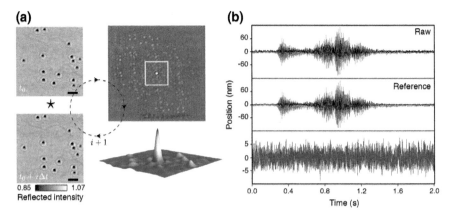

Fig. 3.11 Concept of self-referencing via image cross-correlation. a Workflow diagram to track the position of the entire image, which serves as the reference signal. First, an image taken at t_0, the kernel, is cross-correlated ($\star$) with a test image taken at $t + i\Delta t$, where Δt corresponds to the measurement frame time. This produces an intensity map with a maximum corresponding to the total displacement between the two images. Next, the total displacement is determined by finding the centre of mass of the maximum, thus providing a single point in the positional time-series. Finally, this process is repeated for all images within a dataset. **b** Representative example of the position time-series of a particle in (**a**) before and after self-referencing

setup, a more versatile and cost-effective alternative can be implemented by self-referencing.

In the concept of self-referencing, the constant scattering background signal is used as a reference from which the position of the whole image can be extracted as a function of time, much like the case for isolated diffraction limited spots. Under the assumption that most of the single-molecule features only constitute a small fraction of the total image, topographical or spatial changes due these features play no significant role in the time trace obtained by tracking the entire image. The whole images were tracked by using the principle of 2D cross-correlation [15]:

$$C(x, y) = \sum_i \sum_j I(x + i, y + j) K(i, j) \tag{3.6}$$

where I corresponds to the test image taken at time t, and K to the reference one, also referred to as the kernel and typically assigned with the position at time t_0 (Fig. 3.11a). The product of the cross-correlation function is an intensity map, with a maximum representing the best alignment between both images. The position of the maximum with respect to the centre of the image corresponds to the displacement of the image with the interval $t - t_0$. The centre of the mass of the maximum was determined by the radial symmetry algorithm [10].

There are two major consequences for self-referencing in iSCAT. Firstly, the position of the whole image can be used as a feedback parameter to provide lateral stabilisation of the sample stage [16]. Secondly and more importantly, all forms of extrinsic noise due to lateral drift or vibrations can be removed in the post-acquisition phase without compromising the SNR of the single-particle tracks as shown in Fig. 3.11b. This means that for most SPT applications, stabilisation of the sample area is an unnecessary process and will likely do more harm than good, since the response time will be limited by the electronics and mechanical elements of the feedback loop. Here it is important to stress that the self-referenced time trace has the same temporal resolution as the acquisition and thus can deal with high-frequency noise, a feature that no other method can accomplish (Fig. 3.11b).

References

1. Capitanio, M., Cicchi, R., Pavone, F.S.: Position control and optical manipulation for nanotechnology applications. Eur. Phys. J. B-Condens. Matter Complex Syst. **46**, 1–8 (2005)
2. Janesick, J.R. Photon Transfer. SPIE, 1000 20th Street, Bellingham, WA 98227-0010 USA (2007)
3. Cheung, S., Kamath, C.: Robust background subtraction with foreground validation for urban traffic video. EURASIP J. Appl. Signal Process. (2005)
4. Panahi, S., Sheikhi, S., Hadadan, S.: Evaluation of background subtraction methods. In: Techniques and Applications, Digital Image Computing (2008)
5. Parks, D.H., Fels, S.S.: Evaluation of background subtraction algorithms with post-processing. In: IEEE 5th International Conference on Advanced Video and Signal Based Surveillance (2008)
6. Small, A.R., Parthasarathy, R.: Superresolution localization methods. Annu. Rev. Phys. Chem. (2013)
7. Kechkar, A., Nair, D., Heilemann, M., Choquet, D., Sibarita, J.B.: Real-time analysis and visualization for single-molecule based super-resolution microscopy. PLoS ONE **8**, e62918 (2013)
8. Neubeck, A., Van Gool, L.: Efficient non-maximum suppression. In: ICPR 2006: Proceedings of the 18th International Conference on Pattern Recognition, vol. 3, pp. 850–855 (2006)
9. Wolter, S., et al.: rapidSTORM: accurate, fast open-source software for localization microscopy. Nat. Methods **9**, 1040–1041 (2012)
10. Parthasarathy, R.: Rapid, accurate particle tracking by calculation of radial symmetry centers. Nat. Methods **9**, 724–726 (2012)
11. Jaqaman, K., et al.: Robust single-particle tracking in live-cell time-lapse sequences. Nat. Methods **5**, 695–702 (2008)
12. Jonker, R., Volgenant, T.: Improving the Hungarian assignment algorithm. Oper. Res. Lett. **5**, 171–175 (1986)
13. Jonker, R., Volgenant, A.: A shortest augmenting path algorithm for dense and sparse linear assignment problems. Computing **38**, 325–340 (1987)
14. Chu, P.C., Beasley, J.E.: A genetic algorithm for the generalised assignment problem. Comput. Oper. Res. **24**, 17–23 (1997)
15. Gelles, J., Schnapp, B.J., Sheetz, M.P.: Tracking kinesin-driven movements with nanometre-scale precision. Nature **331**, 450–453 (1988)
16. Mantooth, B.A., Donhauser, Z.J., Kelly, K.F., Weiss, P.S.: Cross-correlation image tracking for drift correction and adsorbate analysis. Rev. Sci. Instrum. **73**, 313–317 (2002)

Chapter 4
Anomalous Diffusion Due to Interleaflet Coupling and Molecular Pinning

The full chapter was written by myself but parts have been adapted from the following publication: Spillane, K. M.*, Ortega Arroyo, J.*, de Wit, G., Eggeling, C., Ewers, H., Wallace, M.W and Kukura, P. Interleaflet coupling and molecular pinning causes anomalous diffusion in bilayer membranes. *Nano Lett.* **14,** 5390-5397 (2014) [1], *:authors contributed equally and are copyright (2014) of the American Chemical Society. Both the experimental work and the data analysis were performed by Dr. Katelyn Spillane, Gabrielle de Wit and myself. Specifically, measurements with 20 and 40 nm AuNPs, sub-trajectory data analysis, localisation precision assessments and implementation of the high-speed data acquisition triggering were performed by myself; the controls and receptor titration experiments with 40 nm AuNPs were performed by Dr. Spillane; and receptor titration with 20 nm AuNPs and corresponding sub-trajectory analysis of that data-set were performed by Gabrielle de Wit. Dr. Ewers provided the DO-GM1 lipid and Dr. Eggeling measured the diffusion coefficient using fluorescence correlation spectroscopy.

4.1 Introduction

The plasma membrane is a highly heterogeneous system composed of a mixture of lipids and proteins distributed amongst two leaflets. Although the plasma membrane, an integral part of the cell, participates in a wide range of functions, its main role is to relay information between the interior and exterior of the cell. This information can be in the form either of molecules, lipids, proteins, and even organelles, as in the case of endocytosis/exocytosis; or forces, as in the case of chemotaxis [2] or cell adhesion. However the underlying mechanisms involved in the information exchange process are poorly understood, especially for events such as a receptor binding to an extracellular leaflet lipid, or rearrangement of membrane components in the absence

© Springer International Publishing AG 2018
J. Ortega Arroyo, *Investigation of Nanoscopic Dynamics and Potentials by Interferometric Scattering Microscopy*, Springer Theses, https://doi.org/10.1007/978-3-319-77095-6_4

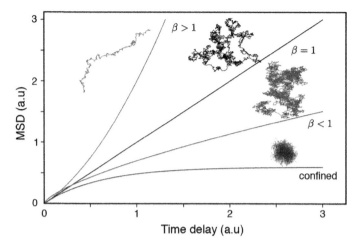

Fig. 4.1 Classification of the types of motion by the mean-squared displacement. Representative trajectories and corresponding MSD curves for the different types of motion a single molecule can exhibit: super-diffusive (orange), normal Brownian (black), anomalous sub-diffusive (blue) and confined (purple)

of transmembrane proteins [3]. Several mechanisms have been proposed [4] based on line tension [3, 5, 6], direct interaction of lipids across the bilayer [7, 8], molecular pinning [9], and the transient formation of cholesterol-rich, ordered membrane domains [10]. All these mechanisms have in common the involvement of characteristic spatio-temporal distributions of the components of the membrane. As a consequence of this organisation, the mobility of individual proteins and lipids deviate from Brownian motion, thus making them a target for single-particle tracking studies.

Deviations in Brownian motion can be classified according to the time-dependent variations in the distribution of particle displacements; a property that is captured by the mean-squared displacement metric (MSD, Fig. 4.1) [11]. When the distribution contains much larger displacements than expected from random motion the observed behaviour is classified as anomalous super-diffusion [12]. Levy walks [13], ballistic motion [14], and the processive dynamics of motor proteins like kinesin, dynein and myosin are all examples of anomalous super-diffusion. Conversely when the distribution contains, on average, smaller displacements the motion is classified as anomalous sub-diffusion [15].

One of the shortcomings of such a general classification is that processes as different as transient binding due to short-lived lipid-lipid or lipid-protein interactions, presence of obstacles [16], and spatial-dependent mobilities due to different viscosities [17], all lead to motion categorised as anomalous sub-diffusion. Furthermore such a method of classification is susceptible to experimental artefacts associated with: large labels [18], poor localisation precision [19], and the time resolution of the measurement [20]. As a consequence of this and the difference in time- and length-scales probed by different single-molecule techniques, several inconsistent and even contradictory results exist across the field of membrane biophysics such as the model of hop-diffusion in cell membrane. Super-resolved fluorescence correlation spectroscopy [21] and high-speed SPT of 40 nm AuNP labels with 15 nm localisation precision achieved by darkfield [22] both provided evidence of anomalous diffusion of lipids in live cell membranes in the form of hop-diffusion. However SPT performed with molecular-sized fluorescent dyes as labels did not show any signs of anomalous behaviour down to the 300 μs time-scale [23]. Furthermore alternative explanations for the observed anomalous diffusion in cells have been proposed, for instance the presence of local nanoscopic membrane roughness [24].

Thus, classification of the mobility of a membrane component as anomalous provides little meaning beyond deviation from a purely random behaviour, unless mechanistic insight is provided, and the potential artefacts associated with a label are addressed. For that reason, in this Chapter I present the results obtained from a SPT assay on a model membrane system where the lipid and substrate interactions were controlled and characterised. Importantly, by using interferometric scattering microscopy we were able to probe several orders of magnitude in time with sub-nm localisation precision.

Given the complexity of cellular membranes and the large number of uncontrolled variables, we opted for a bottom-up approach based on a minimal one-component supported lipid bilayer (SLB) system and characterised the motion of its components under different substrates [25]. These bilayers were composed primarily of 1,2-dioleoyl-sn-glycero-3-phosphocholine (DOPC) lipids and doped at concentrations ranging from 0.03–1.00% molar with a GM1 bovine brain ganglioside (GM1) receptor. The choice for GM1 is based on its involvement in lipid clustering and lipid-lipid interactions which has been observed experimentally in the form of anomalous diffusion in cells [10, 26], phase segregation induced by cholera-toxin [27], and partitioning into liquid-ordered domains [28].

4.2 Experimental Methods

4.2.1 Materials

The compounds 1,2-dioleoyl-sn-glycero-3-phosphocholine (DOPC), 1,2-dihexadecanoyl-sn-glycero-3-phosphoethanolamine-N-(cap biotinyl) (DPPE), and GM1 bovine brain ganglioside (GM1) were purchased from Avanti Polar Lipids (Alabaster, AL). DO-GM1 was prepared as described before [6]. Biotin-labelled cholera toxin B (CTxB) subunits from *Vibrio cholera* were purchased from Sigma-Aldrich (Milwaukee, WI) and reconstituted with water to give a solution containing 50 mM Tris buffer, pH 7.5, 200 mM NaCl, 3 mM NaN₃, and 1 mM sodium EDTA. Gold nanoparticles functionalised with streptavidin were purchased from British Biocell International (Cardiff, U.K.), diluted to a concentration of 9×10^{10} particles/ml and incubated with a ten-fold excess of biotin-CTxB at room temperature for one hour. Excess CTxB was removed by centrifuging the AuNP/CTxB sample for 2 min at 14000 g and resuspending the pellet in bilayer buffer (10 mM HEPES, pH 6.8, 200 mM NaCl and 2 mM $CaCl_2$).

4.2.2   Vesicle Preparation

Small unilamellar vesicles (SUVs) were prepared by the vesicle extrusion method. Lipids in organic solvent were mixed in a glass vial and the solvent evaporated first under a gentle stream of nitrogen for 5 min and then under vacuum for 30 min. The dried lipid film was resuspended to a lipid concentration of 1 mg/mL in bilayer buffer. Lipid suspensions were vortexed for 1 min, hydrated at room temperature for 30 min and then passed 21 times through a 100 nm polycarbonate membrane using a mini extruder (Avanti Polar Lipids), resulting in clear suspensions of SUVs. SUVs were stored at 4 °C and used within 24 h.

4.2.3   Substrate Preparation

No. 1.5 borosilicate coverslips (Menzel-Glaser, Braunschweig, Germany) were etched in 2:1 H_2O_2:HCl for 10 min, followed by thorough rinsing with ultrapure water (Merck Millipore, Billerica, MA). The clean substrates were dried with a gentle stream of nitrogen and etched with oxygen plasma for 8 min at 50 W power immediately prior to vesicle deposition (Diener Electronic, Plasma System Femto). Mica substrates were prepared by bonding a 22 mm square sheet of mica (Agar Scientific, Essex, U.K.) to a clean coverslip using optical adhesive (Norland Optical Adhesive 61). Immediately prior to vesicle deposition, the mica was cleaved leaving a thin, optically transparent layer adhered to the cover glass.

4.2.4   Supported Lipid Bilayer Formation

Sample chambers were assembled by placing a CultureWell silicon gasket (Grace Bio-Laboratories, Bend, OR) onto the glass or mica substrate. SLBs were formed by adding 20 µl bilayer buffer followed by 10 µl of the 1 mg/ml SUV suspension to the 30 µl sample well and incubating for 5 min. Excess SUVs were rinsed away with 3 ml bilayer buffer and then 2.5 µl of AuNP/CTxB or AuNP/streptavidin solution was deposited and allowed to incubate for 5 min. Excess particles were rinsed away with 1 ml bilayer buffer prior to imaging.

4.2.5   Instrument Setup Parameters

The magnification for high-speed imaging was set to 100× which provides an effective pixel size of 106 nm. Depending on the size of the AuNP tracked the incident power was varied between 20–30 kW/cm^2 and 2–3 kW/cm^2 for 20 and 40 nm particles, respectively. In the case of tracking 40 nm particles this is equivalent to focusing approximately 0.3 mW onto a 2×2 μm^2 spot size at the sample.

4.3   Experimental Results

4.3.1   GM1 Undergoes Anomalous Diffusion in Supported Lipid Bilayers

The assay was optimised under wide-field-of-view (30×30 μm^2) iSCAT imaging, achieved by confocal beam scanning. Formation of the bilayer and removal of excess vesicles was assessed prior to addition of the AuNP/CTxB conjugates. CTxB was conjugated to the AuNPs through a biotin-streptavidin linker, and it was estimated that a AuNP has approximately 25 and 40 bound CTxB, for 20 and 40 nm sized gold particles, respectively. Non-specifically bound AuNP/CTxB conjugates were flushed away and resulted in images such as Fig. 4.2a. The signal from 40 nm AuNP/CTxB conjugates bound to GM1 was on the order of 55%, for 20 nm AuNP/CTxB conjugates was 9%, and those from excess vesicles ranged between 1–5%. The excess vesicles typically had a density of one every 5×5 μm^2 and a fraction of them were mobile, thus causing specificity issues in the detection of smaller sized nanoparticle labels (<20 nm). Thus for the remainder of this chapter, most of the results are reported for 40 AuNP labels due to their unequivocal distinction from mobile vesicles.

High-speed imaging was performed by underfilling the objective with a collimated beam at $\lambda = 532$ nm producing a field of view of approximately 4.0 μm^2. The localisation precision in the high-speed imaging channel was assessed by recording the relative motion of two immobile particles on a coverslip at the lowest exposure time setting of our camera, 10 μs, and thus fastest frame time, 20 μs. Assuming the particles are indeed immobile, the fluctuations in the inter-particle distance serve as a metric for the maximum achievable localisation precision at the incident power used. For the incident power typical of the experiment with 40 nm AuNP, a localisation precision of 1.3 nm was achieved; however this was not a lower bound, as an increase of the intensity to 30–40 kW/cm^2 for the same particle pair achieved sub-nm precision (Fig. 4.2b). This is in stark contrast to state-of-the-art fluorescence microscopy which can only achieve 1 nm precision at either limited time resolutions or observation times. Similarly, dark-field microscopy results of 40 nm AuNPs used as labels for individual lipids in a membrane at 40 kHz frame rate (25 μs exposure time) only

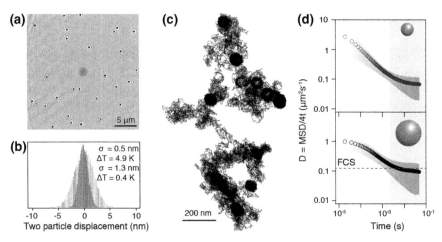

Fig. 4.2 Anomalous sub-diffusion of GM1 in supported lipid bilayers on glass substrates. a Representative interferometric scattering image of 40 nm AuNP/CTxB conjugates diffusing on a supported lipid bilayer. Shaded green region indicates the relative size difference between the wide-field-of-view and high-speed imaging modalities of iSCAT performed in these experiments. Scale bar: 5 μm. **b** Corresponding interparticle separation fluctuations between two 40 nm AuNP/CTxB conjugates after subtraction of the average value for two different incident illumination intensities. The localisation precision is assessed as the spread in the fluctuations and calculated as $\sigma/2^{1/2}$. **c** Sample trajectory of 40 nm AuNP/CTxB/GM1 on glass recorded at 50,000 frames/s and with more than 2×10^5 data points. Scale bar: 200 nm. **d** Time-dependent mobilities for 20 and 40 nm AuNP/CTxB/GM1 on glass. Open circles represent the mean taken of all acquired trajectories, and the red lines the standard deviation resulting from inhomogeneous broadening. Grey shaded out regions denote the typical time-scales probed by fluorescence SPT experiments. Dotted green line: macroscopic diffusion coefficient measured by fluorescence correlation spectroscopy. Number of data points per trajectory: 1×10^5

reported 15 nm localisation precision [22]. This shows that iSCAT measurements decouple time resolution from localisation precision.

Due to the fast movement of the AuNP/receptor complexes and the small field of view in the high-speed imaging channel, the data on receptor mobility was acquired by an automated custom trigger-sequence operating in real-time. This automation removes selection bias, introduced by user-triggering, and a higher throughput, which is limited by the slower response time of a person. The triggering sequence operates by polling every millisecond whether a subset of frames (50 images per ms) contains a diffraction limited spot that resembles the AuNP/receptor complex in question. The polling relies on the local non-maximum suppression algorithm [29] combined with signal contrast thresholding. Briefly, the algorithm divides an image into smaller sections called blocks and detects the local maxima of each block. Each local maxima is compared against a range of user-defined contrasts. If the local maxima falls within the range then a candidate is found. If a candidate is detected within a distance of 800 nm away from the centre of the image, the data acquisition trigger is initialised; otherwise it just keeps polling. Once the trigger is initialised the program streams a fixed amount of frames into the RAM before the program polls again for a designated event. If the trigger is still on, frames are continuously streamed into RAM. This processes stops when either the maximum pre-allocated RAM space is reached or the event trigger stops, which occurs when a particle leaves the illumination area. The data is then stored to the hard-drive and the polling resumes. If the data is stored to the hard-drive, a cool-down time of 20 s is assigned so that the computer can re-synchronise with the camera frame rate, as the action of writing to the hard-drive causes the computer to lag behind the camera acquisition.

This trigger scheme acquires the longest possible trajectories and does so in the most memory efficient manner. With memory efficiency defined as the fraction of frames that contain a particle diffusing within the illumination area per total amount of frames acquired. Trajectories were reconstructed based on the greedy algorithm, described in Chap. 3, from the centre of mass positions of the label as a function of time. Despite the potential artefacts associated with the greedy algorithm, the low density of the AuNP/receptor complex below 1 μm^2 ensured these were kept to a minimum. Examination of the diffusion of 40 nm AuNP/CTxB bound to GM1 in a SLB made from DOPC doped with 0.03 mol % GM1 on a plasma-cleaned glass substrate acquired at 50 kHz, revealed circular nanoscopic regions of transient confinement (Fig. 4.2c); a characteristic feature of anomalous diffusion.

To quantitatively examine the time dependence of the mobility of the complex, we calculated the mean-square displacement [30] as a function of the time lag as:

$$\text{MSD}(\Delta t_n) = \frac{1}{M-n-1} \sum_{j=1}^{M-n-1} [\mathbf{r}(\Delta t_{j+n}) - \mathbf{r}(\Delta t_j)]^2 \tag{4.1}$$

where M is the total number of frames in the trajectory, n is a positive integer that determines the time interval, and $\mathbf{r}$ is the particle displacement during time interval $\Delta t_n = n\Delta t$ with $\Delta t = 20$ μs referring to the experimental frame time. Using the relationship between MSD and two dimensional diffusion coefficient (D):

$$\text{MSD} \sim 4D\Delta t^\beta, \tag{4.2}$$

the motion of the particle was classified as anomalous diffusion if the anomalous coefficient had a non-unity value, $\beta \neq 1$. For convenience, we evaluated the time-dependent mobility as the logarithmic plot of the diffusion coefficient ($D = \text{MSD}/4\Delta t$) versus time, which has a ($\beta - 1$) slope. This means that a zero slope indicates Brownian diffusion, while negative and positive slopes represent anomalous sub- and super-diffusion, respectively.

Because these measurements encompassed time-scales over four orders of magnitude, the transition from anomalous to Brownian diffusion was determined from a single trajectory as opposed to several measurements at different exposure times and with different particles [31]. Furthermore the macroscopic diffusion coefficient measured independently by fluorescence correlation spectroscopy (0.14 $\pm$

$0.01 \, \mu m^2 \, s^{-1}$) agreed with the range of values ($0.09 \, \mu m^2 \, s^{-1}$) obtained at time-scales longer than 10 ms in our experiments [26].

According to this analysis, both 20 and 40 nm AuNP/CTxB conjugates bound to GM1 in a SLB formed on plasma-cleaned glass exhibited anomalous sub-diffusive motion over three orders of magnitude in time, from 20 µs to 10 ms time scale (Fig. 4.2d). After tens of ms the motion transitions into Brownian diffusion resulting from extensive averaging and indicating that CTxB-bound GM1 diffusion is confined on short (<10 ms) time-scales. iSCAT experiments with 20 nm particles resulted in similar trajectories and time-dependent mobilities with minor differences for sub-ms time lags.

4.3.2 Transient Confinement Causes Anomalous Diffusion

Rather than just classifying the motion as anomalous, we investigated the dynamics responsible for deviations in Brownian diffusion, represented in this case by the periods of transient confinement. These transient dynamics were mainly characterised by two populations: ring-like and Gaussian-like structures (Fig. 4.3a), each with a confinement area of approximately 30 nm in diameter. Other types of confinement areas were observed such as elliptical, half-ring and hybrid shapes but their occurrence was far less frequent. These confinement events were detected by a sub-trajectory analysis routine based on the assumption that the average particle displacement over a given time is far smaller for a transient binding event than the value predicted by a purely random diffusion. Here, a value of $1 \, \mu m^2 \, s^{-1}$ based on the measured time dependent mobility within the first 100 µs of Fig. 4.2a was taken as the nominal diffusion coefficient of the AuNP/receptor complex. To minimise false positives and given the transition between anomalous and Brownian diffusion in the 10 ms time-scale, the bandwidth was lowered to 100 Hz (by averaging) and the total displacement was calculated as:

$$|\mathbf{r}(t + \delta t) - \mathbf{r}(t)|. \tag{4.3}$$

Pairwise distance values less than 50 nm were classified as confined, as opposed to the 200 nm displacements expected for random motion at the diffusion coefficient specified. Although a more robust and parameter-free likelihood change-point detection algorithm [32–34] could have been performed at the full detection bandwidth to identify very short-lived transient events, the high computational cost became prohibitive.

To gain further insight into the nature of the two populations of confinement events, Gaussian and ring-like, the spatial probability density map for each case was determined by overlapping all the transient binding data points (Fig. 4.5b). Gaussian-like distributions in a spatial probability map arise from dynamics described by a freely diffusing particle within a harmonic potential [35]. Given our experiment, it is unlikely that the model membrane contains areas of localised harmonic potentials;

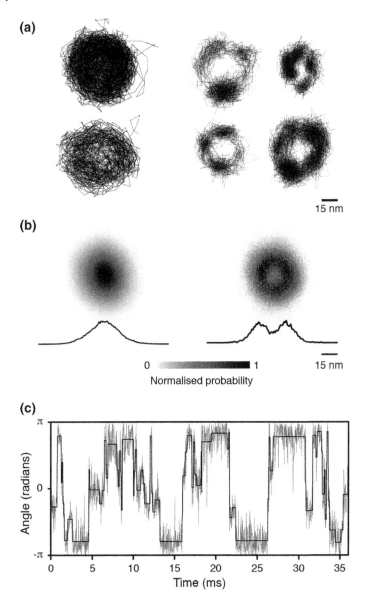

Fig. 4.3 Analysis of transient binding events. a Examples of individual transient confinement events for 40 nm AuNP/GM1 complexes for the two main populations observed: Gaussian and ring-like. **b** Spatial probability maps and corresponding cross-sections for the Gaussian-like (228 events) and ring-like (93 events) populations at 0.03% GM1 on glass. Scale bars: 15 nm. **c** Representative time-trace of a ring-like event showing several binding sites in polar coordinates, only the angle is show for clarity. Thin black line: average angular position of the binding sites detected by change-point analysis

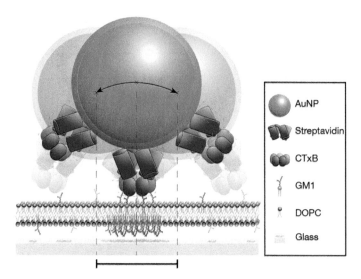

Fig. 4.4 Transient binding model for CTxB in supported lipid bilayers. Schematic of the tracking assay during a transient binding event drawn to scale for a 20 nm AuNP. The curved arrow indicates the achievable nanoscopic rocking motion of the label and the dashed lines the achievable fluctuations in the centre of mass of the label even for a completely immobilised CTxB

instead this observation can be explained by tethered-particle-motion [36] between the gold label and a tether described by GM1 headgroup/CTxB/biotin/streptavidin linker (Fig. 4.4). In this case the harmonic potential is defined by the inherent flexibility of the linker between GM1 and the gold particle. As a result of this description the Gaussian-like population is extracted from the ensemble by selecting the confinement events whose radial distribution can be precisely described by a Rayleigh distribution ($R^2 > 0.95$), which is nothing more than a descriptor of a two dimensional Gaussian distribution in polar coordinates.

Confinement events that do not satisfy these criteria were classified as ring-like. Other shapes were excluded by the criteria that ring-like distributions have small variations in the distance away from the centre of mass. Specifically this was implemented by only selecting events whose fluctuations in the radial distribution were at least 2.2 standard deviation units smaller than the average confinement radius. This value was based on the observed linker flexibility, localisation precision and size of the confined area. Within the ring-like structures we often observed well-defined binding sites that were frequently revisited (Fig. 4.3c). These observations may be explained by a model where multiple interactions between more than one CTxB from the AuNP/CTxB conjugate and diffusing GM1 receptors occur (Fig. 4.4).

4.3.3 Concentration Dependent Dynamics of Transient Binding

Assuming a uniform distribution of GM1, the receptor density was estimated at $1/(40 \times 40)$ nm^2 for the lowest concentration where specific binding of AuNP/CTxB conjugates was observed (0.03% molar); at this point multiple GM1 interactions with the same AuNP/CTxB complex should be rare. Therefore concentration dependent dynamics were investigated by varying the concentrations of nt-GM1 and characterised by radial distribution analysis of all transient binding events. Briefly, all the confinements events were overlapped and the ensemble radial probability density was calculated and then fit to a Rayleigh distribution. Deviations from the Rayleigh distribution, characterised by the residuals in the fit, provided a qualitative assessment of the degree of GM1 interactions given that a tethered particle motion model is defined by the properties of a single GM1 headgroup/CTxB/biotin/Streptavidin linker (Fig. 4.5a). At the lowest concentration the radial density was well characterised by the Rayleigh distribution but as the concentration increased, so did the deviations, as shown by the residuals trace. Despite the same occurrence in Gaussian-like confinement events, the average confinement size defined by the FWHM of the Gaussian distribution, increased linearly yet with a slow rate of change with GM1 concentration (Fig. 4.5b).

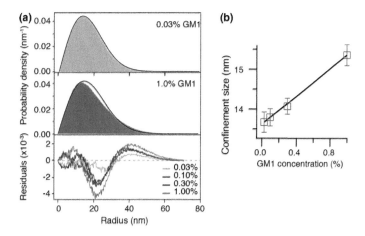

Fig. 4.5 Concentration dependent dynamics of GM1 immobilisation. a Top panel: Radial probability density plots and corresponding fits to a Rayleigh distribution for all confinement events recorded at 0.03% and 1.0% nt-GM1 concentrations, respectively. Total number of data points for each distribution: 2×10^6. Bottom panel: Concentration dependent plot of the residuals in the fit to a Rayleigh distribution. **b** Average change in confinement size as a function of nt-GM1 concentration for Gaussian-like confinement events. The confinement size is evaluated as the full-width-at-half maximum. Error bars denote the standard deviation in the average confinement size

4.3.4   Recovery of Brownian Motion upon Tuning Substrate Interactions and Interleaflet Coupling

In the experiments described above the substrates, in this case glass coverslips, were treated by oxygen plasma cleaning, a process known to functionalise the surface with hydroxyl groups, increase the surface roughness and even induce membrane defects [26, 37]; all of which could serve as potential mechanisms for the transient binding dynamics. To determine the origin of the observed anomalous behaviour and rule out experimental artefacts, such as surface roughness and membrane defects, we repeated the experiments by only modifying one variable at a time. Specifically, we tuned the membrane-substrate interactions and the potential acyl-chain mediated interleaflet coupling.

To investigate the role of membrane-substrate interactions, we tuned the interactions by using chemically inert mica instead of plasma-cleaned glass coverslips. Under these conditions all trajectories, summing up to more than 5×10^5 data points, lacked transient confinement events and were significantly shorter than those using a plasma-cleaned glass as the membrane substrate (Fig. 4.6). Analysis of the time-dependent mobilities of the AuNP/GM1 complexes indicated Brownian motion for all time-scales longer than 100 μs, in agreement with the shorter acquired trajectories. At time-scales shorter than <100 μs, anomalous super-diffusion was detected, suggesting the presence of directional motion or the presence of a flow; nevertheless

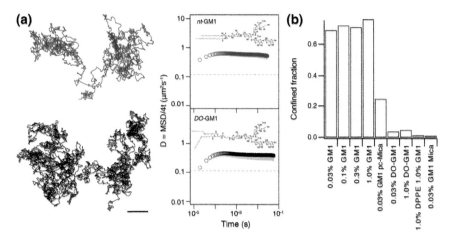

Fig. 4.6 Tuning the degree of transient confinement. a Sample trajectories and time-dependent mobilities of 40 nm AuNP/CTxB/nt-GM1 on mica (red) and 40 nm AuNP/CTxB/DO-GM1 on plasma cleaned glass (black). Blue shaded region on each molecular structure of GM1 analogues denotes the common sugar head group. Data points per trajectory: 2×10^4. Scale bars: 100 nm. **b** Confined fraction as a measure of transient confinement and anomalous diffusion for different GM1 concentrations and control experiments: synthetically modified GM1 (DO-GM1) on glass, DPPE with a biotin/streptavidin linker in the presence of nt-GM1 on glass, and nt-GM1 on freshly cleaved and plasma-cleaned mica

such an explanation is quite unlikely for an almost single component membrane. Instead this type of motion can be attributed to "dynamic error" artefacts which are known to occur at early times in MSD analysis whenever measurements are performed at very high localisation precision, which in our case ranged from 1.5–0.5 nm [20]. The source of "dynamic error" artefacts is the finite exposure time of the camera, which averages the position of the particle and results in an underestimation of the mean squared particle displacement. The recovery of Brownian motion in experiments on mica as a substrate suggest that substrate interactions with the lower leaflet are necessary for the occurrence of transient confinement. For instance, the hydroxyl groups on the glass coverslips could immobilise the hydroxyl-containing GM1 head groups in the lower leaflet. Nevertheless the effect from substrate roughness or plasma cleaning-induced defects could not be ruled out.

To investigate the role of acyl-chain mediated interleaflet coupling [8] between GM1 groups, the sphingosine base comprised of fully saturated hydrocarbon chains of the naturally occurring GM1 (nt-GM1) was replaced with the unsaturated chains from glycerophospholipid 1,2-dioleoyl-sn-glycero-3-phosphoethanolamine (DOPE), to produce a GM1 analogue termed DO-GM1. The measured time-dependent mobilities of DO-GM1 on plasma-cleaned glass substrates using 40 nm AuNPs mimicked the results from the mica substrate, that is Brownian diffusion and apparent superdiffusion on similar time-scales. Nevertheless, the average mobility was slightly slower than mica, and on the level of individual trajectories rare and short-lived transient binding events were detected. The observation of Brownian motion, together with AFM measurements on surface roughness (RMS = 0.6 nm), exclude surface roughness as the source of anomalous diffusion, given that the substrate-membrane interactions are the same as in the analogous nt-GM1 experiment. Furthermore, the presence of unsaturated chains in DO-GM1, and with it a weakened interleaflet coupling, drastically decreased the frequency of transient events, thus suggesting the need for acyl-chain mediated interleaflet coupling to transmit the origin of transient immobilisation to the upper leaflet, where the AuNP label is diffusing.

To confirm the simultaneous requirement of interleaflet coupling and lower leaflet-substrate interactions for transient binding and its nanoscopic origin, three further control experiments were performed. Firstly, experiments were performed to test whether the interactions between the substrate and the sugar-head group from GM1 played a role. To do so, biotinylated DPPE lipids in DOPC bilayers with streptavidin-functionalised AuNPs on plasma-cleaned glass substrates in the absence and presence of nt-GM1 (1.0% DPPE, 1.0% GM1) were tracked. These DPPE lipids possess saturated chains just like GM1 that could potentially lead to strong interleaflet interactions. Secondly, the effect of DO-GM1 concentration on plasma-cleaned coverslips was investigated (0.03 and 1% DO-GM1). In both cases the recovery of Brownian motion was observed at all times-scales longer than 100 μs. Thirdly, upon plasma treatment of chemically inert mica, anomalous diffusion in the form of transient binding was observed for AuNP/nt-GM1 complexes.

As a more robust metric for anomalous diffusion, we summarised the above experiments with respect to the fraction of time a GM1/AuNP complex was transiently confined, defined in this thesis as confined fraction (Fig. 4.6b). The mean confined fraction for all GM1 concentrations measured (0.03–1.00 mol %) for nt-GM1/Gold complex on a plasma-cleaned glass substrate was 0.72 ± 0.03 with zero representing no confinement and one representing complete immobilisation. Thus, the confined fraction metric provided a more accurate description of the underlying dynamics compared to the time-dependent mobilities, given that the latter only reports ensemble behaviour and as a result overlooks rare events such as the transient immobilisations in the DO-GM1 assays on plasma-cleaned glass.

These result confirm that interleaflet coupling between GM1 in the lower and upper leaflet combined with membrane-substrate interactions, specifically between the GM1 head groups in the lower leaflet and the substrate, are sufficient to cause transient confinement and thus anomalous diffusion dynamics in these model membrane systems. Furthermore, the very weak dependence of the confined fraction on the concentration of nt-GM1 for plasma-cleaned glass experiments, suggests that the origin of transient immobilisation is not the GM1 itself but rather an intrinsic property of the surface functionalisation. This property could likely be attributed to the density of hydroxyl pinning sites on the surface caused by plasma-cleaning.

4.4 Discussion

4.4.1 Importance of Simultaneous Localisation Precision and Time Resolution

Much of our understanding of membrane structure and dynamics at the single-molecule levels has been the result of fluorescence and purely scattering-based imaging measurements. Nevertheless these approaches are insufficient to simultaneously probe the relevant time- and length-scales for motion at the level of individual membrane components; either because of a limited photon flux or the potential to perturb the underlying dynamics. The results in this chapter demonstrate the ability of interferometric scattering microscopy to follow the motion of individual lipids and receptors with simultaneous nanometre localisation precision and tens of microsecond temporal resolution; thus making it an ideal method for single-molecule membrane biophysics studies.

As a consequence of this spatio-temporal decoupling it becomes possible to: investigate nanoscopic dynamics inaccessible to other methods, reinterpret previous conclusions, and re-assess the importance of localisation precision. For instance, consider the scenario in which the measurements presented in this chapter were performed with 15 nm rather than 1 nm localisation precision (Fig. 4.7). Although the motion at the ensemble level would still be classified as anomalous diffusion, the underlying dynamics would not correspond to transient binding. Instead, the regions of transient confinement increase in size and overlap with the rest of the trajectory thus giving the impression of corralled motion. Furthermore a hop-diffusion analysis identifies different sized compartments within the trajectory, analogous to SPT

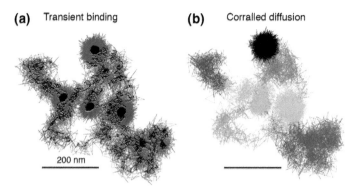

Fig. 4.7 Effect of localisation precision on transient binding. a Example trace of a 40 nm nt-GM1/AuNP complex diffusing on a plasma-cleaned glass supported lipid bilayer. Black: original trace taken with 1.2 nm localisation precision. Red: same trace but with 15 nm Gaussian noise added. **b** Same trace as in (**a**) subjected to hop-diffusion analysis different colours highlighting the different compartments identified. Imaging rate: 50,000 frames/s. Data points: 1×10^5

experiments on single lipids labelled with 40 nm AuNP. These observations of hop-diffusion in cells led to the hypothesis that the plasma membrane of some cells is compartmentalised by the underlying cytoskeleton meshwork [31, 38]. Thus as the result of a poor localisation precision, we have arrived at an entirely different interpretation; from one based on transient immobilisation due to GM1 interactions with the substrate to another based on a compartmentalised membrane.

4.4.2 Thermal and Optical Force Considerations

We used the results from a previous simulation based on the paraxial beam approximation to estimate the total forces acting on a 20 nm gold particle [39]. Briefly, the authors simulated the forces acting on a 80 nm gold particle for a 1 mW total beam power with $\lambda = 532$ nm focused to a spot by an objective lens with NA $= 0.9$ ($\approx$290 kW/cm^2). Given the beam profile and incident intensity, the range of forces varied between 0 and 1.2 pN, with the maximum occurring when the particle is at the beam focus (See Fig. 4) [39]. Note, that the forces acting on the gold particle can be decomposed into a gradient and a scattering force, which scale with Re $\{\alpha\}$ or Im $\{\alpha\}$, where α denotes the complex polarisability and Re and Im refer to the real and imaginary parts, respectively [40]. The complex polarisability is described by:

$$\alpha = \epsilon_m \pi \frac{D^3}{2} \frac{\epsilon_p - \epsilon_m}{\epsilon_p + 2\epsilon_m} \tag{4.4}$$

With this expression, and using the results of the simulation, we estimated the magnitude of the total forces acting on a gold particle of 20 nm in diameter to be

$4^3 = 64\times$ smaller than for a 80 nm one. Since our experimental intensities were $10\times$ lower than those used in the simulation the maximum optical force that a 20 nm gold particle experiences was approximately 1.9 fN (1.2 pN / 640). Similarly for a 40 nm gold particle, although the forces would only be $2^3 = 8\times$ smaller, the intensities used in most experiments were $100\times$ lower than those reported in the simulation, resulting in total optical force on the order of 1.5 fN (1.2 pN / 800). We can therefore conclude that irrespective of the particle size and at light intensities used in these experiments, the optical forces exerted by the beam are on the order of several femtonewtons, and only result in temperature increments of <2 K at the surface of the particle, which are much too small to significantly affect the diffusion on the length-scales studied in this work.

4.4.3   Membrane Defects, Labelling Artefacts, and CTxB Induced Aggregation Do Not Cause Transient Binding

Defects in the membrane, caused by buffer washes, substrate interactions and even induced by specific membrane components, are one of the most common sources for anomalous diffusion in supported lipid bilayers [41]. However with two experimental observations reported in this chapter we can completely rule out this possibility. Firstly, if defects were responsible for anomalous diffusion, both DO-GM1 and nt-GM1 would have exhibited similar diffusive properties, which is in stark contrast to the experimental observations (Fig. 4.6a). Secondly, even under the assumption that differences in DO-GM1 and nt-GM1 lipid properties may cause a different propensity for defect formation, these defects would have also been present during the tracking measurements of DPPE in an nt-GM1-doped SLB, and yet again no anomalous diffusion was observed (Fig. 4.6a).

With regards to potential labelling artefacts, the Brownian motion recovered from tracking DPPE in GM1-doped SLBs suggests that the particle cross-linking or label-membrane interactions are unlikely causes for anomalous diffusions. If the AuNP labels indeed caused transient immobilisation, anomalous sub-diffusion would have been observed in all negative control experiments (Fig. 4.6b). We also rule out the possibility of non-specific binding of the gold nanoparticle to the membrane as a potential source, as AuNP without CTxB did not interact with GM1-containing SLBs and were easily washed away upon buffer rinses.

Furthermore comparison of the results obtained using inert and plasma-cleaned mica substrates excludes phase separation and nanoscopic aggregation due to free CTxB in solution as a possible cause for the observed transient confinement events.

4.4.4 Transient Binding Requires Substrate Interaction and Interleaflet Coupling

The recovery of Brownian motion upon changing the acyl chains from nt-GM1 to DO-GM1, and tuning the substrate interactions of nt-GM1 from plasma-cleaned glass to mica, indicates that only the combination of nt-GM1 with a plasma-treated surface leads to transient binding. Because our experiments are performed on an effectively single component system, lipid composition or substrate-induced asymmetry of the lipid distribution have little to no influence. Moreover, a recent study on the leaflet distribution of GM1 in DOPC bilayers formed on UV/ozone-treated silica showed that 85% of the receptors were located on the upper leaflet and the content of GM1 followed a linear relationship with concentration in the range between 0–5% molar [42]. This effect was mainly caused by the charge in the lipid headgroup of GM1. Therefore, both nt-GM1 and DO-GM1 on glass should exhibit the same distribution between the two leaflets given that both experiments were performed on identical, plasma-cleaned glass substrates. Thus, our results suggest that two conditions are required for anomalous diffusion in this model membrane system: (a) GM1 molecules in the lower leaflet must be immobilised through interactions with the surface and (b) the immobilised molecules must interact with CTxB-cross-linked GM1 molecules in the upper leaflet through the hydrophobic core of the bilayer via long, straight aliphatic chains.

4.4.5 Multiple CTxB-GM1 Interactions Result in Ring-Like Structures

Our ability to super-resolve the spatiotemporal dynamics of the label during periods of transient binding provides us with information about the nanoscopic origin of the observed anomalous diffusion. To illustrate this, it is helpful to consider the molecular details of the tracking assay under the scenario of a single CTxB interaction between an immobilised GM1 and the gold label. When drawn to scale, the distance between the GM1 headgroup and the centre of mass of the AuNP label is non-negligible and contains three main linking elements, each possessing a certain degree of flexibility: GM1-CTxB , CTxB-streptavidin and streptavidin-AuNP (Fig. 4.4). Simple geometric arguments suggest that nanoscopic movement analogous to tethered particle motion induces variations in the centre of mass on the order of 15 nm for both 20 and 40 nm AuNPs. In addition, such movement when viewed over tens of milliseconds is described by 2D Gaussian spatial distribution around the centre of mass as observed experimentally for one population of transient binding events (Fig. 4.3a, b).

The population of ring-like transient confinement events can not be described by the above explanation. However, addition of a second interaction between CTxB from the same nanoparticle with another possibly diffusing GM1, yields motion centred around the immobilised central CTxB with a radius comparable to the maximum motion of the centre of mass of the label. Despite the relatively weak GM1/CTxB dissociation constant, $K_d \approx 10^{-8}$, this multiple bound hypothesis agrees with the ring-like structures observed (Fig. 4.3a).

Moreover as the likelihood of multiple CTxB/GM1 interactions increases with GM1 concentration, the radial probability density plot representative of all confinement events is expected to further deviate from a Rayleigh distribution, characteristic for single tethered particle motion and single particle-interactions. Such a result is evidenced by the concentration dependence of the deviation between the radial probability density plots and the Rayleigh fit (Fig. 4.5a). Even for Gaussian-like confinement events, the increase in nearby GM1 concentration should lead to an increase in the apparent confined area due to transient interactions with additional GM1, even if they do not result in cross-linking (Fig. 4.5b). These interactions are possible because on average the label spends more time near the membrane surface as a result of the nanoscopic rocking caused by the inherent flexibility of the GM1/AuNP linkers [43].

4.4.6   A Model of Transient Binding: Molecular Pinning

The tight lateral confinement on the <20-nm scale and the observation of ring-like confinement events are both consistent with transient immobilisation of a single membrane-bound CTxB subunit on the nanometre scale. Given that transient binding requires both native GM1 lipid tail domains and a plasma-cleaned substrate, we propose that the most likely origin for immobilisation are hydrogen bonding interactions between small patches of surface hydroxyl groups caused by plasma treatment and the hydroxyl groups from the sugar-head groups of GM1. These interactions cause clustering of GM1 on the lower leaflet.

Although a single pinned GM1 molecule on the lower leaflet could be responsible for the immobilisation of five CTxB-bound GM1 molecules in the upper leaflet, such a scenario is unlikely. We therefore, hypothesise that the distribution of GM1 in the lower leaflet is characterised by numerous <10 nm clusters with high GM1 content, in contrast to a uniform distribution. These GM1-enriched clusters with higher viscosity communicate with the upper leaflet through acyl-mediate interleaflet interactions, leading to transient (<10 ms) nanoscopic confinement events. This is supported by AFM measurements of comparable assays, however whether the aggregation occurs in the lower or upper leaflet of the bilayer was not determined [44]. Moreover our general mechanism agrees with early observations of interleaflet coupling [45], although these required the presence of macroscopic domains.

4.5 Conclusion and Outlook

This chapter shows that even in a minimal membrane model system, transient binding can occur in the absence of transmembrane components and thereby change the mobility of single proteins bound to receptors. In this particular case, we found that a combination of proximal GM1 interactions with a plasma-treated substrate and GM1 hydrocarbon chain saturation were sufficient to induce transient immobilisation and anomalous sub-diffusion on the sub-millisecond time scale. Analysis of the dynamics involved in the confinement events suggests that confinement occurs on the <10 nm scale, consistent with temporary immobilisation, i.e. molecular pinning, of GM1-bound CTxB. More importantly, we have elucidated an alternative mechanism, based on interleaflet lipid-lipid interactions, for information exchange across a membrane bilayer in the absence of any transmembrane components.

There are many prospects for future work to further understand the origins of nanoscopic immobilisation and anomalous diffusion. For instance, patterned substrates with specific functional groups would allow specific tuning of the strength and density of membrane/substrate interactions. Asymmetric bilayers produced by Lagmuir-Blodgett/Langmuir-Schaefer [46, 47] could be used to explore the mechanisms of interleaflet coupling and the effect of lipid composition asymmetry, a characteristic property of plasma cell membranes. The use of lipid headgroups or scattering labels with varying valency would determine how many cross-linked GM1 molecules are required for transient binding. Finally, one could envision combining label-free studies of nanoscopic lipid domains with SPT to investigate the role of lipid phase separation.

References

1. Spillane, K.M., et al.: High-speed single-particle tracking of GM1 in model membranes reveals anomalous diffusion due to interleaflet coupling and molecular pinning. Nano Lett. **14**, 5390–5397 (2014)
2. Sourjik, V.: Receptor clustering and signal processing in E. coli chemotaxis. Trends Microbiol. **12**, 569–576 (2004)
3. Liu, J., Sun, Y., Drubin, D.G., Oster, G.F.: The mechanochemistry of endocytosis. Plos Biol. **7**, e1000204 (2009)
4. Klotzsch, E., et al.: Conformational distribution of surface-adsorbed fibronectin molecules explored by single molecule localization microscopy. Biomater. Sci. **2**, 883–892 (2014)
5. Baumgart, T., Hess, S.T., Webb, W.W.: Imaging coexisting fluid domains in biomembrane models coupling curvature and line tension. Nature **425**, 821–824 (2003)
6. Ewers, H., et al.: GM1 structure determines SV40-induced membrane invagination and infection. Nat. Cell Biol. **12**, 11–18 (2009)
7. Wan, C., Kiessling, V., Tamm, L.K.: Coupling of cholesterol-rich lipid phases in asymmetric bilayers. Biochemistry **47**, 2190–2198 (2008)
8. Collins, M.D., Keller, S.L.: Tuning lipid mixtures to induce or suppress domain formation across leaflets of unsupported asymmetric bilayers. Proc. Natl. Acad. Sci. U.S.A. **105**, 124–128 (2008)

9. Honigmann, A., et al.: A lipid bound actin meshwork organizes liquid phase separation in model membranes. eLife **3**, e01671 (2014)
10. Eggeling, C., et al.: Direct observation of the nanoscale dynamics of membrane lipids in a living cell. Nature **457**, 1159–1162 (2009)
11. Metzler, R., Jeon, J.-H., Cherstvy, A.G., Barkai, E.: Anomalous diffusion models and their properties: non-stationarity, non-ergodicity, and ageing at the centenary of single particle tracking. Phys. Chem. Chem. Phys. **16**, 24128–24164 (2014)
12. Saxton, M.J.: Single particle tracking. In: Fundamental Concepts in Biophysics (2009)
13. Yu, C., Guan, J., Chen, K., Bae, S.C., Granick, S.: Single-molecule observation of long jumps in polymer adsorption. ACS Nano **7**, 9735–9742 (2013)
14. Huang, R., et al.: Direct observation of the full transition from ballistic to diffusive Brownian motion in a liquid. Nat. Phys. **7**, 576–580 (2011)
15. Sokolov, I.M.: Models of anomalous diffusion in crowded environments. Soft Matter **8**, 9043–9052 (2012)
16. Tsai, J., Sun, E., Gao, Y., Hone, J.C., Kam, L.C.: Non-brownian diffusion of membrane molecules in nanopatterned supported lipid bilayers. Nano Lett. **8**, 425–430 (2008)
17. Pinaud, F., et al.: Dynamic partitioning of a glycosyl-phosphatidylinositol-anchored protein in glycosphingolipid-rich microdomains imaged by single-quantum dot tracking. Traffic **10**, 691–712 (2009)
18. Mascalchi, P., Haanappel, E., Carayon, K., Mazères, S., Salomé, L.: Probing the influence of the particle in single particle tracking measurements of lipid diffusion. Soft Matter **8**, 4462 (2012)
19. Martin, D.S., Forstner, M.B., Käs, J.A.: Apparent subdiffusion inherent to single particle tracking. Biophys. J. **83**, 2109–2117 (2002)
20. Savin, T., Doyle, P.S.: Static and dynamic errors in particle tracking microrheology. Biophys. J. **88**, 623–638 (2005)
21. Wawrezinieck, L., Rigneault, H., Marguet, D., Lenne, P.-F.: Fluorescence correlation spectroscopy diffusion laws to probe the submicron cell membrane organization. Biophys. J. **89**, 4029–4042 (2005)
22. Fujiwara, T.K., et al.: Phospholipids undergo hop diffusion in compartmentalized cell membrane. J. Cell. Biol. **157**, 1071–1082 (2002)
23. Wieser, S., Moertelmaier, M., Fuertbauer, E., Stockinger, H., Schütz, G.J.: (Un)Confined diffusion of CD59 in the plasma membrane determined by high-resolution single molecule microscopy. Biophys. J. **92**, 3719–3728 (2007)
24. Adler, J., Shevchuk, A.I., Novak, P., Korchev, Y.E., Parmryd, I.: Plasma membrane topography and interpretation of single-particle tracks. Nat. Methods **7**, 170–171 (2010)
25. Sackmann, E.: Supported membranes: scientific and practical applications. Science **271**, 43 (1996)
26. Mueller, V., et al.: STED nanoscopy reveals molecular details of cholesterol- and cytoskeleton-modulated lipid interactions in living cells. Biophys. J. **101**, 1651–1660 (2011)
27. Lingwood, D., Ries, J., Schwille, P., Simons, K.: Plasma membranes are poised for activation of raft phase coalescence at physiological temperature. Proc. Natl. Acad. Sci. U.S.A. **105**, 10005–10010 (2008)
28. Dietrich, C., Volovyk, Z.N., Levi, M., Thompson, N.L., Jacobson, K.: Partitioning of Thy-1, GM1, and cross-linked phospholipid analogs into lipid rafts reconstituted in supported model membrane monolayers. Proc. Natl. Acad. Sci. USA **98**, 10642–10647 (2001)
29. Neubeck, A., Van Gool, L.: Efficient non-maximum suppression. In: ICPR 2006: Proceedings of the 18th International Conference on Pattern Recognition, vol. 3, pp. 850–855 (2006)
30. Saxton, M.J., Jacobson, K.: Single-particle tracking: applications to membrane dynamics. Annu. Rev. Biophys. Biomol. Struct. **26**, 373–399 (1997)
31. Kusumi, A., et al.: Paradigm shift of the plasma membrane concept from the two-dimensional continuum fluid to the partitioned fluid: high-speed single-molecule tracking of membrane molecules. Annu. Rev. Biophys. Biomol. Struct. **34**, 351–378 (2005)

32. Horvath, L.: The maximum likelihood method for testing changes in the parameters of normal observations. Ann. Statist. **21**, 671–680 (1993)
33. Chen, J., Gupta, A.K.: Testing and locating variance changepoints with application to stock prices. J. Am. Stat. Assoc. **92**, 739–747 (1997)
34. Horvath, L., Steinebach, J.: Testing for changes in the mean or variance of a stochastic process under weak invariance. J. Stat. Plan. Inference **91**, 365–376 (2000)
35. Jin, S., Haggie, P.M., Verkman, A.S.: Single-Particle tracking of membrane protein diffusion in a potential: simulation, detection, and application to confined diffusion of CFTR Cl channels. Biophys. J. **93**, 1079–1088 (2007)
36. Lindner, M., Nir, G., Vivante, A., Young, I.T., Garini, Y.: Dynamic analysis of a diffusing particle in a trapping potential. Phys. Rev. E **87**, 022716 (2013)
37. DeRosa, R.L., Schader, P.A., Shelby, J.E.: Hydrophilic nature of silicate glass surfaces as a function of exposure condition. J. Non-Cryst. Solids **331**, 32–40 (2003)
38. Kusumi, A., Shirai, Y.M., Koyama-Honda, I., Suzuki, K.G.N., Fujiwara, T.K.: Hierarchical organization of the plasma membrane: Investigations by single-molecule tracking vs. fluorescence correlation spectroscopy. FEBS Lett. **584**, 1814–1823 (2010)
39. Fedoruk, M., Lutich, A.A., Feldmann, J.: Subdiffraction-limited milling by an optically driven single gold nanoparticle. ACS Nano **5**, 7377–7382 (2011)
40. Arias-Gonzalez, J.R., Nieto-Vesperinas, M.: Optical forces on small particles: attractive and repulsive nature and plasmon-resonance conditions. J. Opt. Soc. Am. A **20**, 1201–1209 (2003)
41. Sharonov, A., et al.: Lipid diffusion from single molecules of a labeled protein undergoing dynamic association with giant unilamellar vesicles and supported bilayers. Langmuir **24**, 844–850 (2008)
42. Carton, I., Malinina, L., Richter, R.P.: Dynamic modulation of the glycosphingolipid content in supported lipid bilayers by glycolipid transfer protein. Biophys. J. **99**, 2947–2956 (2010)
43. Windisch, B., Bray, D., Duke, T.: Balls and chains–a mesoscopic approach to tethered protein domains. Biophys. J. **91**, 2383–2392 (2006)
44. Shi, J., et al.: GM 1clustering inhibits cholera toxin binding in supported phospholipid membranes. J. Am. Chem. Soc. **129**, 5954–5961 (2011)
45. Seul, M., Subramaniam, S., McConnell, H.M.: Monolayers and bilayers of phospholipids at interfaces: interlayer coupling and phase stability. J. Phys. Chem. **89**, 3592–3595 (1985)
46. Moraille, P., Badia, A.: Nanoscale stripe patterns in phospholipid bilayers formed by the langmuir-blodgett technique. Langmuir **19**, 8041–8049 (2003)
47. Picas, L., Suárez-Germà, C., Teresa Montero, M., Hernández-Borrell, J.: Force spectroscopy study of langmuir-blodgett asymmetric bilayers of phosphatidylethanolamine and phosphatidylglycerol. J. Phys. Chem. B **114**, 3543–3549 (2010)

Chapter 5
Structural Dynamics of Myosin 5a

The full chapter was written by myself but parts have been adapted from the following publication: Andrecka, J.*, Ortega Arroyo, J.*, de Wit, G., Fineberg, A., MacKinnon, L., Young, G., Takagi, Y., Sellers, J.R. & Kukura, P. Structural dynamics of myosin 5 during processive motion revealed by interferometric scattering microscopy. *eLife*. **4**, e05413 (2015) [1] *: authors contributed equally, and are copyright (2015) of eLife Science Publications. Both the experimental work and the data analysis was performed by Dr. Joanna Andrecka, Gabrielle de Wit, Lachlan MacKinnon, Adam Fineberg, Gavin Young and myself. Specifically, the data acquisition measurements with 20 nm AuNPs at different frame rates, sub-trajectory data analysis, kinetic analysis, user friendly trajectory visualisation and analysis software implementation were performed by myself. High speed data acquisition at 10,000 frames/s and correlative blue and red iSCAT traces were acquired by both Dr. Andrecka and myself. Dr. Andrecka collected and analysed the correlative GFP fluorescence and iSCAT, ATP and label size dependence assays. The doubly labelled experiments were performed by Gabrielle de Wit and Dr. Andrecka. Lachlan MacKinnon, Adam Fineberg and Gavin Young quantified the statistics of single actin filament switching and directionality of the transient state. Dr. Yasuharu Takagi and Dr. James R Sellers provided the myosin V constructs and critical unpublished data in the form of electron microscope image and processivity kinetics.

5.1 Introduction

Molecular motors are a set of protein complexes responsible for most of the types of motion and the generation of force at the cellular level [2]. A family of these motors known as cytoskeletal due to their association with actin and microtubule

© Springer International Publishing AG 2018
J. Ortega Arroyo, *Investigation of Nanoscopic Dynamics and Potentials by Interferometric Scattering Microscopy*, Springer Theses,
https://doi.org/10.1007/978-3-319-77095-6_5

filaments, undergo large-scale conformational changes that are fuelled by chemical energy in the form of adenosine triphosphate, ATP. These conformational changes either result in processive motion of the protein complex or in force generation. Of these cytoskeleton motors, myosin 5a is one of the best understood actin-binding proteins [3] and has been studied extensively by a diverse set of ensemble and single-molecule methods [4–9].

The main function of myosin 5 is to transport cargo towards the barbed (+) end of 7-nm diameter actin filament [4]. It does so by translocating in a hand-over-hand fashion [7] resulting in discrete 74 nm steps of its catalytic sites, known as heads, and 37 nm displacements of its centre of mass [10, 11]. This processive motion is accomplished by a finely tuned kinetic sequence of ATP binding, hydrolysis, phosphate and eventually ADP release, each of which result in Angstrom level changes in the structure of the protein subdomains [12–16]. Although atomic-resolution crystal structures of myosin 5a exist in the presence and absence of the nucleotide, these structures only correspond to the detached state of the acto-myosin complex [17, 18]. With these structures the internally coupled rearrangement of the subdomains leading to phosphate release or dissociation from actin upon ATP binding could be studied in silico; [19–21] thus providing a model for the structural changes potentially induced by binding to actin [22]. Nevertheless experimental observation of these structural dynamics during the processive motion of the motor has been hindered by the requisite of simultaneous localisation precision and temporal resolution, as these changes are likely to be on the Angstrom scale and may be very transient. As a result, the overall stepping behaviour of myosin 5a is well characterised, but the sequence of structural events that leads to the 74 nm step remains poorly understood, namely, how the enzymatic domain converts Angstrom-level structural changes into large scale motion.

Another poorly understood aspect of the dynamics of myosin 5 involves the mechanism by which the heads translocate from one actin binding site to another. This is largely due to the timescale of such motion, as the first passage time of the translocating head is expected to be on the order of 100 μs, [23–25] although the head spends tens of ms in the unbound state given the maximum rates of ATP hydrolysis and binding to actin [23, 26, 27]. From a biophysical perspective, the insight provided by directly following and characterising the type of motion would shed light on the underlying principles of efficient motion at the nanoscale level. Several single-molecule studies aimed at revealing the stepping mechanism have either reported periods of increased flexibility [23, 26, 27] or proposed partitioning of the step into multiple sub-events [23, 28–30]. From these studies the proposed models have in common an initial forward aiming power stroke followed by a Brownian search mechanism of the detached head [23, 31–33]. The power stroke, performed by the attached head, leads to a forward displacement of the pivot point, which in turn facilitates the Brownian rotation of the unbound head. Next, a recovery stroke has been hypothesised to provide the necessary bias and the stability towards the next binding site [34]. However it is debatable how a Brownian-search mechanism could lead to unidirectional stepping motion.

Given the time and length-scales involved in the structural dynamics of myosin 5, any single-molecule experiments intending to probe these events require simultaneous nanometre localisation precision and high temporal resolution without significantly perturbing the system. Here, interferometric scattering microscopy offers an alternative to optical tweezers and single-molecule fluorescence microscopy; as the former satisfies the spatiotemporal constraints, but the use of large probes linked to the motor domain are likely to perturb the dynamics; [35] and the latter introduces minimal perturbations with the use of molecule-sized labels at the expense of either temporal or localisation precision [7]. In this chapter I provide the results obtained from single-molecule studies using iSCAT on a construct of myosin 5a with an N-terminal biotin ligand, which was conjugated with different sized gold nanoparticles, and a C-terminal GFP ligand.

5.2 Experimental Methods

5.2.1 Sample Preparation

Rabbit skeletal muscle actin was prepared as described [36] and stored in liquid nitrogen until used. A 20 μM actin stock solution was prepared in polymerisation buffer (10 mM imidazole, 50 mM KCl, 1 mM $MgCl_2$, 1 mM EGTA, (pH 7.3) containing 1.7 mM DTT, 3 mM ATP). Actin was diluted 20–50 times in motility buffer (MB; 20 mM MOPS pH 7.3, 5 mM $MgCl_2$, 0.1 mM EGTA). Mouse myosin 5a HMM with a C-terminal GFP was expressed in the presence of calmodulin and purified as described [37]. In addition, the N-terminus was modified by the addition of a nucleotide sequence encoding an AviTag peptide (GLNDIFEAQKIEWHE) for site-specific biotinylation with BirA ligase (Avidity). After biotinylation, the sample was aliquoted, flash frozen in 20 μl drops and stored at $-80\,°C$. Before labelling, the myosin sample was diluted in MB containing 40 mM KCl, 5 mM DTT, 0.1 mg ml^{-1} BSA and 5 μM calmodulin.

Gold nanoparticles of 20, 30 and 40 nm in diameter conjugated with streptavidin were purchased from BBI (UK) and directly mixed with biotinylated myosin 5a sample in a 4:1, gold to myosin ratio, consistent with one or zero myosin molecules per gold particle. The mixture was incubated on ice for at least 15 min (sample volume 50 μl, final concentration of myosin 300 pM). The same procedure was used for double gold/Qdot labeling. The quantum dots (emission maximum 565) conjugated with streptavidin were purchased from Invitrogen.

For fluorescence-only imaging, myosin was incubated for 10 min with Atto-647 streptavidin (Atto-TEC). The following oxygen scavenger system was used to increase the fluorescent dye stability: 0.2 mg ml^{-1} glucose oxidase, 0.4% w/v glucose, 0.04 mg ml^{-1} catalase, all purchased from Sigma.

The flow cell was prepared as described [38]. It was first rinsed with 1 mg ml^{-1} solution of poly(ethylene glycol)-poly-l-lysine (PEG-PLL) branch copolymer (Surface Solutions SuSoS, Switzerland) in PBS and incubated for 30 min. Next, it was washed twice with MB before adding the actin solution. After 5 min of incubation, the chamber was washed with MB and the surface was blocked by adding 1 mg ml^{-1} BSA in MB buffer. Finally, the chamber was inspected and myosin-gold conjugate solution containing ATP was added. All assays were performed at room temperature.

5.2.2   Experimental Setup

Interferometric scattering microscopy was performed using AOD confocal beam scanning with a 445 nm laser diode. For two-colour imaging a second diode laser (635 nm) was overlapped with the 445 nm beam path with a dichroic mirror. In the detection arm, the images were separated by an identical optic before being imaged onto two separate CMOS cameras (Photonfocus MV-D1024-160-CL-8) at 333× magnification (31.8 nm/pixel). The incident power was adjusted to 17.9 kWcm^{-2} at the sample to achieve near-saturation of the CMOS camera, which ensured shot noise-limited detection. Fluorescence-only imaging and tracking was achieved with a home-built TIRF microscope using a 635 nm diode laser. The incident power was set to 5 kWcm^{-2} and the fluorescence imaged onto an Andor iXon3 860 EM-CCD camera at 72.1 nm/pixel with a 9.2 μm × 9.2 μm field of view using dielectric filters (Thorlabs) to separate illumination and emission.

5.3   Experimental Results

5.3.1   N-Terminus Labelling Does Not Perturb the Kinetics of Myosin 5a

The motion of the myosin heads was tracked by conjugating streptavidin functionalised gold nanoparticles of 20 nm in diameter to the biotinylated N-terminus domain of the protein. The enhanced light scattering from the gold particle conjugates immersed in an aqueous solution resulted in a decrease in the light intensity of about 8% under illumination at a wavelength of 445 nm (Fig. 5.1a). No differences in processivity or velocity of the myosin-gold conjugates translocating along actin filaments immobilised on the glass substrate were observed (Fig. 5.1b), in agreement with previous single-molecule studies of myosin 5a label-free [39] or labelled at the motor domain, lever arm or stalk [23, 28–30]. Despite the significant size of the label, the point of attachment of the label does not interfere with the mechanochemical cycle.

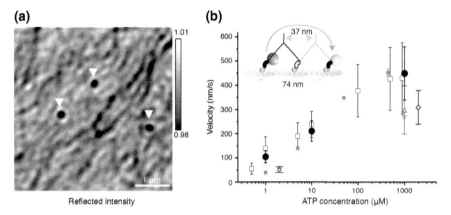

Fig. 5.1 **Myosin 5 activity labelled with 20 nm gold nanoparticles at the N-terminus. a** Representative image of the assay consisting of 20 nm gold particles attached to the N-terminus of myosin as they translocate along actin filaments. Arrows indicate the position of the acto-myosin complexes. **b** Comparison of myosin 5 activity recorded in this work with previous single molecule studies. Black circles: this study. Symbols are defined as: Purple rectangle, [4] Grey squares, [6] Dark blue square, [9] Orange triangle, [40] Pink circle, [41], Red open squares [39]

5.3.2 During Myosin Movement the Motor Domain Undergoes a Transition Between Two Distinct States

Typical traces obtained from tracking myosin-gold conjugates at 100 Hz frame rate and 10 μM ATP concentration moving along actin filaments immobilised on a glass substrate exhibited the characteristic 74 nm stepping pattern. Closer inspection of the recorded trajectories revealed a transition between two distinct states during each step, which are termed state A and state B. Both states are clearly seen in an x-y projection (Fig. 5.2a) and the time trace (Fig. 5.2b). As shown in the lateral trajectory, the transition between these states (AB transition) involves a small, <10 nm, off-axis backward motion of the label attached to the bound head. As a result, the distance from state B to the next state A exceeds that of a 74 nm step; however comparison of transitions between similar states agree with the step size from previous studies (Fig. 5.2c).

An estimate of the localisation precision achieved under these experimental conditions was determined by the positional fluctuations ($\sigma = 0.94$ nm, SD) of a surface-attached label recorded under the same conditions as the trajectory in Fig. 5.2a. The results illustrate simultaneous sub-nm lateral localisation precision of 20 nm gold at 100 Hz (Fig. 5.2b). For actin-bound myosin-gold conjugates the localisation noise increases ($\sigma = 1.6 \pm 0.3$ nm), but remains on average ~2 nm, which allows detection of the AB transitions on the order of 5 nm with a SNR >2 (Fig. 5.2d). The increase in positional fluctuations for gold bound to actomyosin complexes compared to immobilised gold particles can be attributed to a combination of a flexible protein-label linker and the limited ability to completely immobilise actin. Here, actin is only

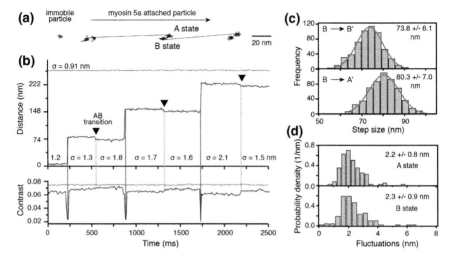

Fig. 5.2 Two states of the motor domain observed during myosin movement. a Sample 2D trajectories for a 20 nm gold particle immobilised on the surface and attached to the N-terminus of myosin 5a. The arrow indicates the direction of myosin movement. **b** The same trajectory depicted as distance travelled vs time and corresponding contrast. Two distinct states (A and B) of the bound head can be clearly observed (black arrows). Standard deviations are given to compare both fixed (blue) and myosin attached particles (black). The brief reduction in iSCAT contrast coincides with 74 nm steps taken by the labelled head (red vertical lines). The behaviour of a non-specifically surface-bound 20 nm gold particle that is completely immobilised is shown for comparison (blue traces in upper portion of panel b). Repeated localisation of the particle throughout suggests a nominal sub-nm lateral localisation precision and a constant scattering contrast. ATP concentration: 10 μM. Scale bar: 20 nm. Imaging speed: 100 frames/s. **c** Step size histograms for post to pre-power stroke and post to post power stroke states. Data was collected at 100 Hz and 10 μM ATP, using 20 nm diameter gold nanoparticles. Number of steps analysed: 554. **d** The positional fluctuations of the A and B states presented as histograms of lateral localisation precision, σ, show no measurable difference in mechanical stability of the two states

immobilised by electrostatic charges provided by the PEG-PLL layer; nevertheless covalent binding with NEM myosin II or α-actinin could provide enhanced sta-bilisation [42]. The imaging speed was limited to 100 Hz to reduce the effects of label-linker flexibility, which increase localisation noise at higher speeds and reduce our ability to identify the AB transition.

5.3.3 The Labelled Motor Domain Moves in Three Dimensions

In addition to the AB transition, a clear drop in the iSCAT contrast was observed whenever the labelled trailing head detached and transitioned to the leading position (Fig. 5.3a, top panel). A similar yet smaller change also occurred during the AB tran-

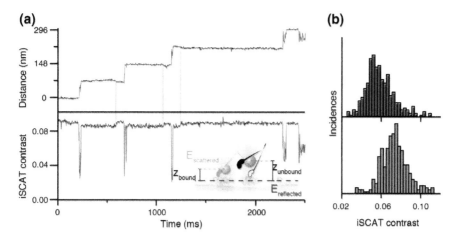

Fig. 5.3 Three-dimensional interferometric tracking of the myosin head. a Distance trace (upper panel) with the simultaneously recorded iSCAT contrast (lower panel). The red subset of the traces corresponds to the unbound head state. Inset: schematic illustrating how the difference in optical path difference for the bound (z-bound) and unbound (z-unbound) state leads to changes in iSCAT contrast due to the interference of the reflected (E-reflected) and scattered (E-scattered) electric fields. **b** Normalised histogram of the average iSCAT contrast while the head is in the transient state (red, N = 329) or bound to actin (black, N = 303). ATP concentration: 10 μM. Imaging speed: 1000 frames/s

sition (Fig. 5.2b, bottom panel). The origin of these contrast changes result from the interferometric nature of the technique, whereby the signal contrast scales according to $\cos \Delta\phi$, where $\Delta\phi$ is the phase difference between scattered and reflected light fields. As the label moves perpendicular to the sample plane, the optical path length and with it the phase difference between the two fields changes by the following relation: $\Delta\phi = 4\pi n z/\lambda$ (Fig. 5.3a, inset). Under the experimental conditions, (aqueous buffer with $n = 1.33$), a complete signal inversion is expected if the label travels $\sim \lambda/4n = 83$ nm away from the surface, making the contrast extremely sensitive to motion perpendicular to the imaging plane.

The variation in iSCAT contrast can therefore be used to determine when the labelled head detaches from actin while tracking a processive myosin 5a (Fig. 5.3a, bottom panel). We often observed these changes in contrast between the actin bound and unbound states and obtained an average change in iSCAT signal during the step (Fig. 5.3b). The drop of 3.5% in iSCAT contrast (40% of the total signal) suggests that the myosin head lifts on average by 24 ± 10 nm from its actin bound position. Although the precision of the measurement was in principle higher, we could not accurately determine the additional phase contributions to the interferometric signal on an individual label basis, which are required for a robust calibration. Overall, the contrast changes with time follow the lateral tracking results in that the labelled head appears to detach only during the 74 nm step; whereas the smaller contrast changes during the AB transition were likely due to a three-dimensional reorientation of the label.

5.3.4   A Conformational Change in the Motor Domain Accompanies the Power Stroke

To investigate the mechanochemical origin of the AB transition, simultaneous tracking of the head and tail domain of myosin 5a was performed. The head domain was labelled with a 20 nm gold particle as previously described, whereas the tail was tagged with a green fluorescent protein (GFP) moiety conjugated to a GFP booster [43]. Given the limited photon flux from a single GFP molecule, which would pose a challenge for single-molecule tracking with <5 nm precision at 100 Hz, [44] the frame rate was reduced to 20 Hz to increase the fluorescence signal. The correlative iSCAT and fluorescence traces showed that whenever the large translocation of the head domain to the next binding site and the small backwards displacement representing the AB transition occurred in the iSCAT channel, they were accompanied by tail translocation events in the fluorescence channel (Fig. 5.4a). This implies that the AB transition corresponds to the pre- to the post-power stroke transition of the attached labelled head (Fig. 5.4b).

If the AB transition corresponds to the pre- to post-power stroke transition, a similar dependence on ATP concentration on the dwell times of both A and B states would be expected, assuming that the label did not affect the stepping kinetics. Thus

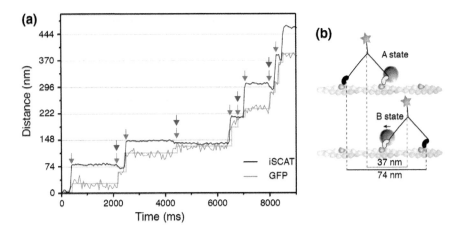

Fig. 5.4   Simultaneous iSCAT and fluorescence tracking of myosin 5a. a Tracking of the scattering signal from the gold nanoparticle attached to the N-terminus (black) and fluorescence signal from the GFP moiety located at the C-terminus of the same myosin 5a molecule (green). Movement of the tail correlates with the labelled head taking its step (74 nm displacement) and with the AB transition (power stroke of the labelled head). Green arrows represent the tail movement, which corresponds to the step of either the labelled or unlabelled head. When the unlabelled head takes its step it coincides with the AB transition within the labelled head (red arrows). Static localisation precisions as previously determined correspond to: 1.6 nm (iSCAT), 8 nm (GFP). **b** Labelling scheme and schematic of the stepping mechanism and the corresponding observables. ATP concentration: 1 μM. Imaging speed for both channels: 20 frames/s

the stepping behaviour of 20 nm gold labelled myosin 5a at different ATP concentrations (1 μM, 10 μM and 1 μM) was recorded and each state separated for dwell-time analysis. To isolate each state, each trajectory was first partitioned into step pairs (Fig. 5.5a). These step pairs contained a single 74 nm translocation together with two of each state, A and B. Differentiation between state A and state B was achieved by analysing the 2D trajectory as a function of elapsed time; whereby a change-point between each state was assigned when the step pair trace showed a spatially separated position with respect to the centre of mass of state A. The corresponding dwell time distributions for both states followed the same trend (Fig. 5.5b): (1) single exponential behaviour at saturating ATP (1 μM) as expected for an ADP release rate-limiting process, (2) single exponential behaviour at 10 μM ATP when the ATP release rate and ADP release rates are both rate-limiting, and (3) bi-exponential behaviour at 1 μM ATP concentration as expected for a process limited by ATP binding [5, 12]. In all cases, the dwell times of each state were identical suggesting that the 20 nm gold particle had no considerable effect on the stepping kinetics.

Using the gold particle as a nanoscopic translocation amplifier, the size of the label was varied to unravel the nature of the structural transition between the pre- to post-power-stroke conformation of the labelled head. The average distance between the centre of mass between each state, defined as the size of the AB transition, increased from 7.4 ± 3.2 nm to 9.5 ± 3.1 nm and 11.5 ± 2.5 nm for 20, 30 and 40 nm diameter labels, respectively (Fig. 5.6a). The corresponding contour plots for the A and B states obtained by superimposing all data points corresponding to detected A states and B states show that the direction of the vector connecting states A and B remains unchanged while the distance increases. This suggests that the AB transition reflects a conformational change in the N-terminus domain in the form of a three-dimensional rotation of the bound head rather than a translation.

The rotational movement of the label can be depicted schematically as in Fig. 5.6b; whereby the red circles mark the positions of two labels in the pre-power stroke (A) state with radii r_1 and r_2, where r_2 is twice the length of r_1 and the grey circles refer to the corresponding post-power stroke (B) states of each label. With d_i representing the distance between the centre of mass (dots) of the labels from the pre- to the post-power stroke conformations, the larger label (r_2) undergoes the same type of motion at the N-terminus but with a larger displacement compared to the smaller label. Applying this minimalistic model to our data, the angle of rotation, α, as well as the distance, x, that defines the origin of rotation from the surface of the protein could be determined following the relations that: $\sin(\alpha/2) = 0.5d_1/(r_1 + x)$ and $d_2/d_1 = (r_2 + x)/(r_1 + x)$. From the hydrodynamic radii obtained by dynamic light scattering for each label (20, 26 and 32 nm), the values of $\alpha = 20°$ and $x = 1.7$ nm were obtained.

Taken together, these observations suggest that a small three-dimensional rotation of the N-terminus domain of the labelled head occurs during the power stroke and coincides when the unlabelled head translocates to the next actin binding site. This motion is then amplified by the size of the label, much like the lever arm, and is

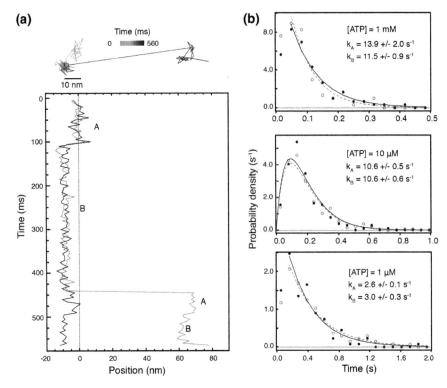

Fig. 5.5 Identification of the pre- and post-power stroke states via a change-point detection algorithm. a Representative distance time trace for each dimension and corresponding 2D-projection of a step pair where the elapsed time has been encoded using a colour gradient. A change-point between state A and B was assigned when a trajectory showed a spatially separated position at a given time point with respect to the centre of mass of state A. **b** Dwell time distributions for pre- and post-power stroke states at different ATP concentrations. In all cases, data and fits for pre-power stroke state are shown in red; data and fits for the post-power stroke state are shown in blue. At saturating (1 μM) ATP concentration (**a**) both dwell time distributions exhibited single exponential behaviour in line with ADP release being the rate-limiting step. The constants for the A state dwell times are given by k_A and that for the B state by k_B. At lower ATP concentrations, sequential ADP release and ATP binding result in bi-exponential behaviour. At 10 μM ATP (**b**), both kinetic constants, k_A, k_B, are almost identical therefore the dwell time distribution is fit to two sequential process with the same rate constant (termed k_A for the dwell time of the A state and k_B for that of the B state. **c** At 1 μM ATP the process is limited by ATP binding. Data taken at 100 frames/s and increased to 400 Hz for 1 μM ATP concentration assays. Total number of steps recorded: 110, 245 and 104, for 1 μM, 10 μM and 1 μM respectively

experimentally reported as the AB transition. A direct consequence of these observations is that the AB transition cannot involve dissociation of the labelled head from actin since it must remain bound while the unlabelled head takes its step, otherwise myosin would detach from actin.

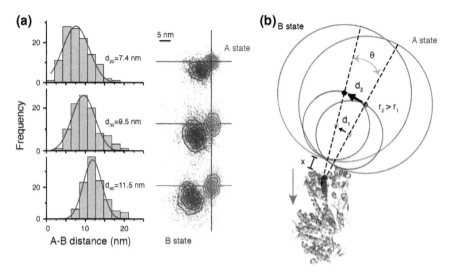

Fig. 5.6 Conformational change within the N-terminal domain during the power stroke. a
Histograms for the distances between A and B states for 20, 30 and 40 nm gold nanoparticle labels
located at the N-terminus and corresponding spatial probability density maps. Total number of steps
recorded: 124, 103 and 116, respectively. **b** Expected movement for two differently-sized labels
attached to myosin 5a during a conformational change in the head domain associated with the
power stroke. The red circle represents the position of the label in the A state, and the grey circle
corresponds to the B state with dots indicating their respective centres of mass. The labels r_1 and r_2
correspond to the radii of both labels where $r_2 = 2r_1$, and d_1 and d_2 correspond to the AB distance
after rotation by an angle θ about an origin located at a distance x from the nanoparticle surface.
The myosin 5a head domain pre-power stroke conformation is shown in orange (PDB: 1W7J). The
lever arm is pointing out (shown in dark blue) and the blue arrow indicates its movement during
the power stroke

5.3.5 Myosin Steps via a Single, Spatially-Constrained Transient State

Upon increasing the time resolution of the experiment by an order of magnitude
to 1 ms (imaging speed = 1000 Hz) the effective localisation precision decreased to
4 nm compared to the measurements at 100 Hz, which can be attributed to the inherent
flexibility of the linker between the label and protein complex. Nevertheless, with this
temporal resolution, periods of increased positional fluctuations of the labelled head
between detachment and reattachment to the next actin binding site were observed,
which had been previously assigned to the Brownian search mechanism (Fig. 5.7a).
These fluctuations when projected laterally in two-dimensions showed a transient
state with a centre of mass just over half way between the two binding sites and
offset by 40 nm perpendicular to the actin filament (Fig. 5.7b).

The dynamics of the motor domain in the unbound state at this time resolution
can be described as a periodic back and forth motion between the observed transient

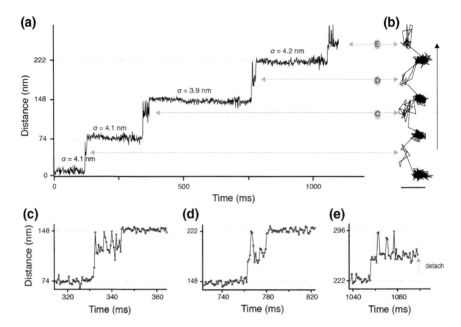

Fig. 5.7 High-speed nanometric tracking of myosin 5 with interferometric scattering (iSCAT) microscopy. a Distance travelled as a function of time for a single myosin 5 molecule biotinylated at the N-terminus and labelled with a 20 nm streptavidin-functionalised gold particle. The lateral localisation precision, σ, defined as the standard deviation of the positional fluctuations of the label while bound to actin is given above each of the actin-attached periods. Inset: schematic of gold-labelled myosin 5 stepping along actin. **b** Corresponding 2D-trajectory with the arrow indicating the direction of movement. **c–e** Close-up of the transient states indicated in (**a**) and (**b**). ATP concentration: 10 μM. Scale bar: 50 nm. Imaging speed: 1000 frames/s

state position and the next actin binding site (Fig. 5.7c, d). A similar behaviour at the end of a processive run was also observed (Fig. 5.7e), although this only occurred when myosin detached when the unlabelled head was bound to actin. The other possible scenario, when myosin detached when the labelled head was bound to actin, was equally measured. To verify that the dynamic features of the motor domain while in the unbound state were not induced by unwanted interactions of the label, a different labelling strategy using streptavidin functionalised with ATTO-647N dyes was used. Traces obtained from the much smaller-sized label exhibited the same characteristic off-axis transient state position and motion of the 20 nm gold nanoparticles (Fig. 5.8a, b). Furthermore, similar dynamics were also displayed for the rare occasions when the leading head detached and reattached without translocating, i.e. repeated back and forth motion and an off-axis transient state (Fig. 5.8c). This suggests that the transient state corresponds to a potential minimum of the one head bound myosin.

The directionality of the transient state with respect to the actin filament remained fixed either to the right or to the left hand side during a translocation event along a

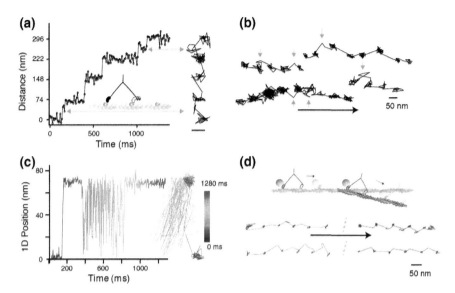

Fig. 5.8 Characterisation of the transient state. a Representative distance time series of a myosin 5 molecule, labelled with an atto-647N/streptavidin conjugate at the N-terminus, tracked using single molecule total internal fluorescence microscopy and corresponding 2D-trajectory. **b** Additional traces exhibiting the transient state marked by arrows. ATP concentration: 10 µM. Scale bar: 50 nm. Imaging speed: 100 frames/s. **c** Representative time trace segment of an event in which the leading head detaches from the actin filament and exhibits the intermediate state behaviour reported in the main text and corresponding 2D-trajectory of the segment. ATP concentration: 1 µM. Scale bar: 50 nm. Imaging speed: 1000 frames/s. **d** Position of the transient state for the same molecule before and after switching actin tracks. ATP concentration: 10 µM. Scale bar: 100 nm. Imaging speed: 100 frames/s

single actin filament (Fig. 5.8d) with a slight preference (66 vs 33%) for right over left-handed walking (from a total of 351 traces). When the myosin moved from one actin filament to another, the directionality of the transient state either switched or remained on the same side with a probability of 40% and 60%, respectively (from a total of 102 events).

To spatially and temporally characterise the transient state as an ensemble, the same approach to isolate the A and the B states was applied on the high-speed trajectories taken at different ATP concentrations. Here, only step pairs with transient states lasting longer than 10 ms were used for the spatial analysis; whereas all step pairs were used for the dwell-time analysis. The dwell time of the transient state followed a single exponential distribution with a lifetime of 17.5 ± 0.6 ms which was independent of ATP (Fig. 5.9a).

Spatial characterisation was achieved by segmenting the tracks into step pairs. Then the centre of mass of the beginning of the step pair was positioned about the coordinate position (0, 0). Tracks were aligned along the horizontal axis by applying a rotation matrix from which the directionality of the transient state was assessed. Finally, all aligned step pair traces were overlapped by flipping them about the actin

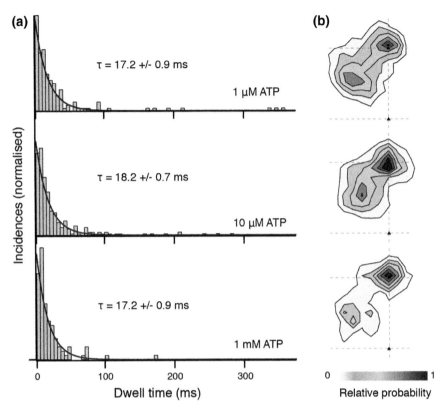

Fig. 5.9 Spatiotemporal dynamics of the transient state as a function of ATP concentration.
a Dwell time histograms at three different ATP concentrations (1 μM, N = 116; 10 μM, N = 223; 1
μM, N = 90). Fitting the dwell time distribution to a single exponential yields an average lifetime
of 17.5 ± 0.6 ms, independent of ATP concentration. **b** Contour maps of the transient state at three
different ATP concentrations. Obtaining traces at high ATP concentration was challenging due to
the small available field of view and rapid detachment of myosin from actin caused by the faster
stepping rate

filament axis (horizontal axis) when necessary to produce a contour map depicting
the probability density of finding the centre of mass of the 20 nm gold particle label
when the motor domain of myosin is in the unbound state during a single step. The
contour map was generated by binning the localisations of each transient state into
18×18 nm² bins. The spatial distribution showed no ATP concentration dependence
within the experimental error (Fig. 5.9b).

A more detailed inspection of tracks containing both the AB transition and the off-
axis transient state showed that the relative location of these events were correlated
(Fig. 5.10a). The resulting overlap of the spatial probability maps of the AB transition
with all transient states taking into consideration the spatial pattern between the two
suggest a possible connection between the directionality of the transient state and the

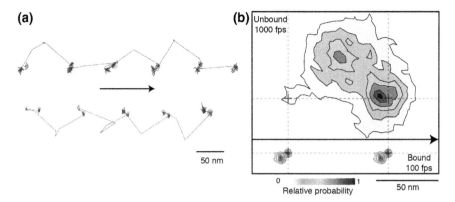

Fig. 5.10 Probability density contour maps of the myosin step. a Simultaneous observation of the AB transition and the transient state imaging at 100 frames/s for right and left handed walking. ATP concentration: $10\,\mu$M. Scale bar: 100 nm. Imaging speed: 100 frames/s. **b** Upper panel represents the transient state of the unbound head. Contour map of a two-dimensional histogram with a $10\times10\,nm^2$ bin width obtained from the 1000 frames/s data (N = 486). Lower panel shows the AB transition within the bound head, a two-dimensional histogram with a $1\times1\,nm^2$ bin width generated using the 100 frames/s data (N = 129). All contributing steps were aligned and those to the right of the filament were mirrored. The arrow represents direction of movement (from left to right)

lever arm movement. Finally, the transient state contour map contains two maxima: one located $\sim$5 nm away of the final actin binding site and the other 40 nm off-axis from the actin filament (Fig. 5.10b).

5.3.6 Transient States Occur on the Same Side of Actin for Each Head Domain

To investigate whether subsequent steps occur on the same or opposite sides of the actin filament, commonly referred to as symmetric and asymmetric hand-over-hand stepping, [45] the two heads of a single molecule were labelled differentially, one with a fluorescent quantum dot and the other with a 20 nm gold particle (Fig. 5.11). The motion of each individual motor domain was followed simultaneously with correlative fluorescence and iSCAT measurements using 473 and 660 nm lasers as the excitation sources, respectively. The chosen excitation scheme provided an off-resonant iSCAT channel, which minimised fluorescence excitation and cross-talk between the channels. To ensure a localisation precision of <5 nm over the time-scale of seconds in the fluorescence channel, the overall imaging speed was lowered to 500 frames/s. The lateral position of the transient state relative to the actin filament coincided across the fluorescence and iSCAT channels for all the recorded events in which a doubly labelled myosin exhibited well defined one head bound states. This observation provided evidence for the symmetric hand-over-hand stepping mechanism.

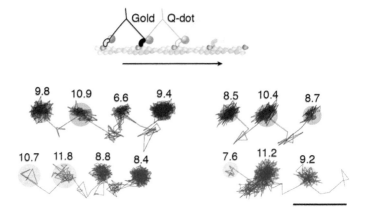

Fig. 5.11 Simultaneous scattering and fluorescence tracking of a single myosin 5a. One head was labelled with a 20 nm streptavidin-functionalized gold particle (red) and the other with a fluorescent quantum dot (blue). Reported values correspond to the standard deviation σ in the position of the bound state (nm) and the shaded regions encompass an area of 3σ. The colours of the traces matches those from the labels in the inset. ATP concentration: 10 μM. Scale bar: 100 nm. Imaging speed: 500 frames/s

5.3.7   The Diffusion Rate of the Unbound Labelled Head Is Comparable to the Frame Time of 1ms

To determine whether similar dynamics were observed at higher time resolution, the tracking experiments were repeated at an imaging speed of 10,000 frames/s (Fig. 5.12a). The presence of a large scattering background in the image due to the presence of gold-myosin complexes in solution limited the localisation precision of these measurements to 9 nm rather than 2 nm. The time traces of transients states lasting longer than 5 ms from 10 trajectories were selected for further analysis ($N = 37$, Fig. 5.12b). Note that the single step pair presented in Fig. 5.12c agreed with the ensemble average transient state probability map from data taken at 1000 frames/s. The autocorrelation function for each isolated position time trace was calculated and later averaged to produce the ensemble autocorrelation function for the transient state (Fig. 5.12d). The diffusive time, a measure of the hydrodynamic environment and thus how long prior information is retained in the system, ($\tau_d = 0.30$ ms) was extracted by fitting a single exponential to the ensemble autocorrelation function and was comparable to the exposure time of 0.56 ms. The fit to a single exponential follows from the assumption that the system can be modelled as a freely diffusing particle in a harmonic potential [46].

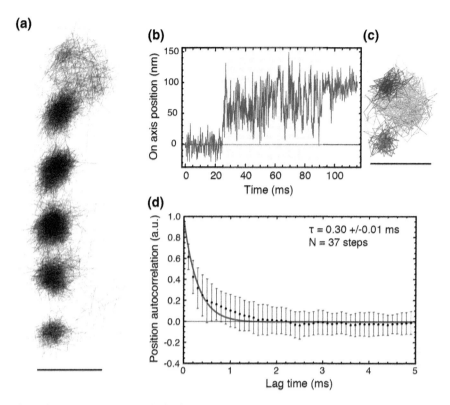

Fig. 5.12 Autocorrelation analysis of the transient state. a High speed 2D trajectory of a 20 nm gold particle attached to the N-terminus of myosin 5a. Imaging speed: 10,000 frames/s. **b** Representative on-axis position time-trace of a step-pair from (**a**). Highlighted area corresponds to the the transient state. **c** Corresponding xy-projection of (**b**) with colours indicating the before (blue), during (green) and after (red) stages of the transient state. **d** Ensemble position autocorrelation function from 37 transient state time-traces. Solid line indicates the best fit to a single exponential function. Scale bars: 50 nm

5.4 Discussion

5.4.1 Association Between an N-Terminus Rotation and the Lever Arm Motion

The tracking results at high localisation precision at 100 frames/s contain several pieces of evidence suggesting that the source of the AB transition is a small underlying structural change in the bound head while the other head performs its step. The observed increase in the magnitude of the AB transition with label size and the small change in contrast are consistent with a rotation of the N-terminal domain by $\sim20°$. Evidence that the labels do not induce any measurable changes in the kinetics is provided by a comparison between the lifetimes of the A and B states (Fig. 5.5b), which represent the dwell time kinetics of stepping for the labelled and unlabelled head. In this case both A and B state lifetimes were the same within experimental error.

Although most SPT labelling strategies opt for the smallest sized marker to minimise perturbing the dynamics, the use of larger sized labels in the tens of nanometres in these experiments, with no detrimental effect to the kinetics, were crucial to reveal the structural dynamics in the N-terminus domain that were otherwise undetectable. Here, the different sized labels amplify minute structural changes occurring in the bound motor domain and convert them into measurable nanometre scale motion. Previous studies using gold labels did not observe this transition likely due to a combination of a different labelling strategy (40 nm or 60 nm gold nanoparticle attached to a calmodulin on the lever arm) and a much lower spatial precision (> 10 nm) [26].

Despite several EM studies failing to detect differences in the position of the SH3 (N-terminal domain) of lead and trail heads, [47–49] results of molecular dynamic simulations of both myosin 5a and myosin 2 based on crystal structures indicate a rotation of the N-terminus during the power stroke transition [17, 19, 21]. These simulation results agree with movement of the N-terminal domain associated with the power stroke taken after the trail head dissociation demonstrated with these experiments.

5.4.2 Sub-steps Along Actin and Leading Head Detachment Do Not Significantly Contribute to the Mechanochemical Cycle

By specifically labelling the N-terminus domain and following the motion of the unbound head during translocation, the dynamics of stepping could be studied in detail with iSCAT. From the 4728 steps composing 635 different tracks not a single sub-step was detected, namely the proposed 64-10 nm stepping pattern. This stepping mechanism was first suggested by a single-molecule fluorescence study [50], where one of the calmodulins bound to the lever arm was fluorescently labelled with a rigidly attached dye. The discrepancy between these two results can be explained by the different labelling strategy. Given that the lever arm moves significantly during the power stroke, whereas the motor domain does not move, the rigidly attached dye experiences an additional displacement which may be interpreted as a sub-step. Furthermore, the orientation of a single dipole can induce systematic errors in determining the position if the PSF is fitted to a 2D Gaussian [51]. Therefore the term 'sub step' is an inappropriate descriptor for the dynamics of the unbound motor domain.

The role the so-called 'foot stomp' event, when the leading head detaches and re-attaches to actin, plays in the mechanochemical cycle of myosin has been an extensive topic of debate in the field ever since its observation [50]. During this first study the frequency of the foot stomp was not indicated; nevertheless a recent

AFM study [52] reported that this event was not only frequent, but also an essential component of the mechanochemical cycle. Contrary to this notion, the sensitivity of our technique to axial displacement showed that both heads of myosin remain firmly bound to actin irrespective of whether the head is leading or trailing (Fig. 5.8c) with the exception being the 74 nm translocation of the head to the next actin binding site.

Although detachment of the leading head was observed its frequency was rare: from all steps taken at 1 and 10 μM ATP and only $\sim$3 and $\sim$0.6% of steps exhibited this behaviour, respectively. At saturating ATP concentration not a single event was detected. Furthermore, the unbinding event was not necessarily followed by reattachment at a different site, suggesting that these events can be explained simply by the binding equilibria between myosin and actin, rather than being an active component of the stepping mechanism. One likely explanation for the increased occurrence of lead head detachment and reattachment in recent AFM studies [52, 53] is the much lower ATP concentration used and the non-negligible interaction of the AFM tip with the protein, which may result in more frequent detachments from actin.

5.4.3 Myosin Preferentially Walks in a Plane Perpendicular to the Glass Surface

Based on the spatial position of the transient state, displaced 40 nm along and perpendicular to the filament, the type of motion of the unbound head can not be explained by a purely rotational diffusion search mechanism. Such a model would require that most measured molecules be bound in the plane parallel to the surface of the glass (Fig. 5.13a). If so, the average transient position of the detached head would be located approximately 56 nm along and no more than 20 nm perpendicular from the filament, as proposed theoretically by a recent simulation [25]. Furthermore a spatial probability map of such rotational diffusion would show one rather than two maxima as reported in this study. Also, the use of BSA to block the surface and prevent non-specific attachment of the gold complexes favours perpendicular binding of the motor as it minimises interaction of the label with the surface; thus making sideways binding to actin (plane parallel to the glass surface) extremely unlikely.

The results of simultaneously tracking the motor domain at 445 and 635 nm illumination provide additional evidence against the rotational diffusion model and its corollary: most molecules are parallel to the glass surface. These correlative iSCAT tracks (Fig. 5.13b) showed that on the occasions when the signal contrast of the particle fell below the detection level in the blue channel, it remained visible in the red channel (Fig. 5.13c, d). In these cases, the off-axis component of the transient state was much smaller compared to those trajectories when the transient state remained visible in both channels, suggesting that these particular myosins were bound parallel to the surface and thus lifted the detached head up by approximately 40 nm.

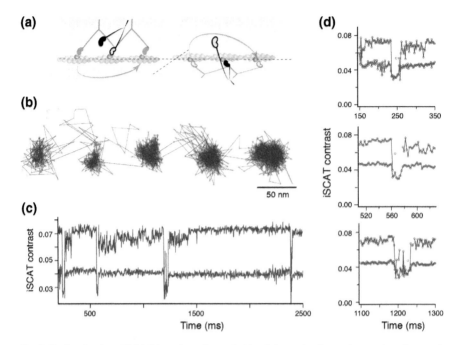

Fig. 5.13 Dual colour iSCAT imaging of myosin 5. a Schematic of myosin stepping along actin with different azimuthal orientations relative to the actin filament. **b** 2D-trajectory of myosin 5 stepping along actin recorded simultaneously with 445 nm (blue trace) and 635 nm (red trace) illumination. Although the latter channel exhibits higher noise due to a lower scattering signal, the transient state is evident in every step even though much of the data are lost in the trace recorded at 445 nm due to the major drop in scattering intensity. **c** Corresponding iSCAT contrast as a function of time. Although the contrast is large in the blue channel, it drops much more significantly during population of the transient state. **d** Zoom-in of the contrast behaviour during the transient state from (**c**). ATP concentration: 10 μM. Scale bar: 50 nm. Imaging speed: 1000 frames/s

Together these observations demonstrate that whenever the spatial position of the transient state occupied an off-axis location with respect to actin and could be measured in the blue channel, the orientation of myosin with respect to the actin filament was perpendicular to the glass surface and thus we can rule out a purely rotational diffusion search mechanism. It is important to emphasise that this data does not contradict the results of a previous dark field experiment, which lead to the rotational diffusion model [26]. The same conclusions can be drawn from our data if the localisation precision matches that of the dark field experiment (17 nm). At such a localisation precision, the directionality of the transient state can no longer be assessed thus leading to a probability density map that is no longer one-sided.

5.4.4 Structurally Constrained Diffusion Leads to Unidirectional Motion

The simultaneous high temporal and spatial precision achieved in this work provided additional insight into the type of motion of the unbound head of myosin to propose a new model for myosin's processivity (Fig. 5.14). While in the two head bound state of myosin 5a, the trailing head detaches from actin upon binding of ATP. The leading head then undergoes a power stroke that leans the protein forward, a motion that exhibits a strong torsional component [40, 41]. After the power stroke, the unbound head arrives at a minimum in the potential energy from where it approaches the next actin binding site in what appears to be one dimensional search along an arc. The transient state lifetime agrees with the duration of increased flexibility reported in optical trapping experiments probing the attachment stiffness at high time-resolution [23] and in single-molecule tracking of the motor domain [26, 27]. The measured lifetime also correlates well with previously reported ensemble rate constant for the weak-to-strong binding transition (47 s^{-1}) [14]. This interpretation is further supported by the presence of an additional density very close to the final binding site, namely the second maxima in the spatial probability density map. This density very likely corresponds to the weakly bound state from which the head frequently moves back to the side position.

The lack of a binding site at the off-axis position of the transient state, 40 nm away from the actin filament, and no reason to pause there pose the question of the origin of such a potential energy minima. However, the structure of the motor suggests a possible explanation, specifically upon considering the intrinsic angle between the lever arms and comparing such a value when the two heads are unbound and bound to actin, i.e. ~37 nm away from each other. Electron micrographs of myosin 5 bound to actin [47] and non-specifically bound to a surface in the absence of actin [54] show clearly peaked angle distributions around 105 and 115°, respectively, suggesting a built-in preference for such a spatial arrangement of the two heads, rather than a truly flexible linkage, another indication that a completely free swivel at the neck-linker is unlikely. By constraining the angle between the lever arms during the translocation, the search space of the motor domain is limited to a one-dimensional rather than a full three-dimensional space. This in turn results in the motor domain preferentially reaching the next binding site.

The position of the transient state together with a possible constrained diffusion suggests the following model where myosin side steps along actin in a combined twisting and leaning motion reminiscent of a compass. Such a mechanism provides the benefit of a built-in bias for the next binding site and ensures that the head only needs to travel along one dimension rather than a lower probability three-dimensional Brownian search. Even if the motor domain fails to bind strongly on first passage, it swings back and forth between the transient state and the desired actin site until tight binding can occur. This suggests that the structure of myosin 5a facilitates and controls the motion of the unbound head to achieve such high specificity in finding the desired binding site.

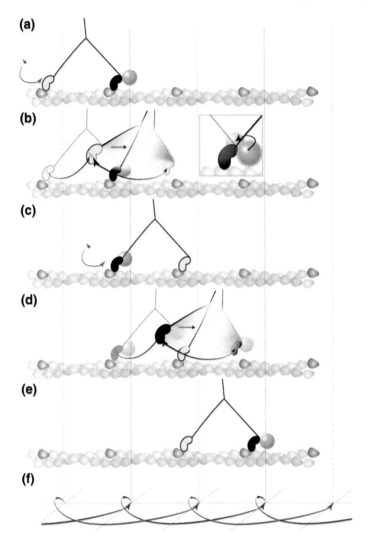

Fig. 5.14 Mechanism of the myosin step. a ATP binding to the trailing head (yellow) and its detachment releases strain stored in the molecule. **b** The bound head (black with a gold nanoparticle attached) performs its power stroke which is accompanied by the AB transition (inset). The labelled head becomes a new trailing head **c** which detaches after ATP binding. **d** It moves forward in a partially twisting motion and occupies an off axis position (transient state). From this position, the head binds the desired binding site while ATP hydrolyses. **e** The step completes as the head binds actin and is repeated by the other head in the same direction dictated by the initial torsional strain. **f** Schematic representation of both heads' movement (lateral projection)

It is important to point out that these conclusions do not reject the presence of a diffusional search mechanism in the translocation dynamics of myosin. The key distinction in this model is that the search space is reduced in dimensionality by structural constraints and in this feature, we propose, lies the efficiency of the processive unidirectional motion. Evidence for the Brownian search mechanism is provided by the measured positional autocorrelation time of 0.3 ms, which shows good agreement with the value predicted theoretically [24]. However, given that most traces were recorded at an exposure time (0.56 ms) comparable to the autocorrelation time, the inherent features of diffusional search were simply missed. However, the location of the transient state, the fact that in several thousand steps a single step with transient localisations on both sides of the actin filament was never observed, together with a predominantly perpendicular orientation of myosin relative to the surface caused by our surface preparation, suggest that diffusion is indeed constrained.

Any label-induced or unwanted steric interactions between the label and the surface as a possible source of this type of motion can be ruled out by the following experimental observations: presence of the transient state for smaller sized labels (Fig. 5.8a, b), changes in contrast during translocation (Fig. 5.3a), lack of non-specific binding by surface passivation, lack of localisations along the actin filament during the one-head bound state (Figs. 5.10b, 5.13b) and the agreement between transient state lifetimes by alternative labelling strategies and single-molecule methods [6, 8, 9].

5.4.5 Relationship Between the Transient State and the AB Transition

The spatial pattern between the transient state position and the AB transition (Fig. 5.10) provides evidence that the AB transition cannot solely be a consequence of a steric interaction between the label and the lever arm. If so, the AB transition would occur only along one direction rather than the two observed, irrespective of whether right- or left-handed walking occurs. Furthermore, although our data demonstrates that the power stroke is associated with the AB transition and the corresponding N-terminus rotation, there is no sufficient experimental evidence at the moment to support a direct causal link between the 20° rotation and the transient state off-axis position. Instead, the off-axis position of the transient state and the direction of the AB transition can be explained by the release of built-up strain on the trailing head upon detachment. This strain is built up initially within the head domain and transmitted towards the lever arm upon each subsequent actin binding event of the motor domain, namely when one of the heads twists before strong attachment to the same actin filament [55].

The proposed hypothesis assumes that the power stroke conformational change in the N-terminal domain measured as the AB transition acts as a strain release reporter. This pre- to post-power stroke conformational change is critical for the rearrangement of the nucleotide binding pocket and/or closure of the internal cleft as it drives the actomyosin energy transduction cycle towards completion. This means that the system can only relax, i.e. the energy stored in the actomyosin complex from the ATPase reaction only becomes available, when the trailing head detaches from actin. An alternative explanation based on the lever arm displacement being responsible for the transient state position can be dismissed upon the observation that the transient state is either to the left or to the right of the filament.

5.4.6 Directionality of the Symmetric Hand-Over-Hand Mechanism

Correlative fluorescence and iSCAT measurements demonstrate that myosin walks in a symmetric hand-over-hand manner where each head swings 180° on the same side of the actin filament, resulting in either a continuous clockwise or counterclockwise rotation. As a corollary, either the cargo rotates together with the motor or there is a swivel within the structure of myosin that releases the built up strain between the cargo and the protein [56]. Evidence for the former has been recently reported but was attributed to thermal motion rather than a symmetric hand-over-hand mechanism, as determined from tracking the position and orientation of a quantum road conjugated to the tail domain of myosin [41].

If myosin walks symmetrically in a hand-over-hand fashion, then the question arises as to what determines the directionality and how this is achieved. Our observations that the directionality of the transient state may change upon binding to a different actin filament rule out the possibility that this decision is an intrinsic property of the protein and instead suggest that the actin filament is involved in the decision making process. Otherwise, the motor would retain its orientation irrespective of changing filament. If actin alone determines the directionality, an equal occurrence of events for each type of motion would be expected, however this was not observed experimentally. Although steric interaction with the surface could induce a preference for a certain directionality due to an initially constrained binding angle, i.e. could be a symmetry breaking element, the predominantly perpendicular orientation of the molecules with respect to the glass surface should be the least affected by this interaction. Although a pure actin filament contribution is very unlikely, our data does not provide any further experimental evidence to reject this hypothesis. This could be addressed by tracking multiple proteins translocating along the same actin filament. In this case, the actin hypothesis can be ruled out if multiple sidedness occurred within a single filament for different molecules.

An alternative explanation can be proposed, where the initial interaction between myosin and actin in the form of the first binding event determines the directionality. This initial binding of the two heads involves a twisting motion, which inherently leads to a build up of torsional strain that is repeatedly stored and released as the protein moves along. This torsional strain in turn leads to the constrained search space and the resulting efficient unidirectional motion previously discussed. It is still unclear how the direction of the initial twist is determined or what causes the symmetry break. One possible option is that the direction of the initial twist is the result of thermal fluctuations, and the symmetric breaking event is simply an effect of the label size or labelling position.

5.5   Conclusion and Outlook

By tracking gold particle labels attached to the N-terminus of myosin 5a with interferometric scattering microscopy at simultaneous high temporal resolution and localisation precision, a conformational change in the N-terminus domain that is associated with the power stroke while the head is bound to actin was directly observed experimentally for the first time. In addition, our results revealed a transient state during the one head bound state, whereby the detached head moved to the next actin binding site via a constrained search space defined by the structure of the molecule. Correlative fluorescence and iSCAT measurements showed both heads of the homodimeric protein follow identical paths along the actin filament, thus confirming that myosin walks symmetrically in a hand-over-hand fashion. More importantly by connecting these results, we provide a mechanism that explains how myosin 5a moves efficiently along the actin network, which may serve as a blueprint for the development of artificial molecular motors and shed new light on the highly efficient mechanochemical cycle of cytoskeletal motors [57–59].

Beyond providing a model for the unidirectional motion of myosin, the results presented in this chapter serve as a proof of concept for at least two future avenues of research and applications of iSCAT towards unravelling structural dynamics of biological systems. On one hand, our approach of using a nanoscale label as an amplifier of conformational changes may become a general tool to study Angstrom-scale structural dynamics of proteins much like FRET [60]. Here the rational design of the labelling scheme will depend on the size and nature of the structural change and the attachment site of the label, thus possibly representing the major experimental challenge. Nevertheless, the use of a single nanoscopic amplifier, as opposed to FRET, offers a higher achievable temporal resolution and reduces the complexity of the labelling route by using a single marker. The use of fewer markers, also avoids the possible scenario when both donor and acceptor molecules are placed within the same domain undergoing conformational change, which would inevitably make the conformational change undetectable. At the moment, only the feasibility of this approach has been demonstrated, but further experiments aimed to quantify its applicability on different systems are needed.

On the other hand, the use of high speed tracking with nanometre-scale precision opens the possibility to perform the equivalent process of super-resolution microscopy on the temporal domain rather than on the spatial one [61, 62]. Analogous to a motion capture suit, by extending these results to different labelling sites along the structure of the myosin and performing correlative tracking with fluorescence a more detailed and accurate picture of the underlying dynamics can be obtained. For instance, the motion of the tail domain and lever arm during the one head bound domain still remains elusive and could be addressed by simply changing the position of the gold particle label. This approach is not restricted to myosin 5 as the same underlying principle governs the processive cytoskeletal motors such as myosin VI and the whole family of kinesins and dyneins.

Finally, the three dimensional tracking capabilities of iSCAT [63] were not fully exploited given lack of a calibration curve and thus the limited control and knowledge over the intrinsic phase constant that is independent of the particle axial position. Future experiments will greatly benefit from a proper characterisation and generalisation of this intrinsic phase beyond specific experimental geometries. The applications are numerous from 3D tracking to the quantification and characterisation of underlying potentials as previously accomplished for fluorescence measurements [64, 65].

References

1. Andrecka, J., et al.: Structural dynamics of myosin 5 during processive motion revealed by interferometric scattering microscopy. eLife **4**, e05413 (2015)
2. Kolomeisky, A.B., Fisher, M.E.: Molecular motors: a theorist's perspective. Annu. Rev. Phys. Chem. **58**, 675–695 (2007)
3. Vale, R.D.: Myosin V motor proteins: marching stepwise towards a mechanism. J. Cell. Biol. **163**, 445–450 (2003)
4. Mehta, A.D., et al.: Myosin-V is a processive actin-based motor. Nature **400**, 590–593 (1999)
5. Rief, M., et al.: Myosin-V stepping kinetics: a molecular model for processivity. Proc. Natl. Acad. Sci. USA **97**, 9482–9486 (2000)
6. Forkey, J.N., Quinlan, M.E., Alexander Shaw, M., Corrie, J.E.T., Goldman, Y.E.: Three-dimensional structural dynamics of myosin V by single-molecule fluorescence polarization. Nature **422**, 399–404 (2003)
7. Yildiz, A., et al.: Myosin V walks hand-over-hand: single fluorophore imaging with 1.5-nm localization. Science **300**, 2061–2065 (2003)
8. Snyder, G.E., Sakamoto, T., Hammer III, J.A., Sellers, J.R., Selvin, P.R.: Nanometer localization of single green fluorescent proteins: evidence that myosin V walks hand-over-hand via telemark configuration. Biophys. J. **87**, 1776–1783 (2004)
9. Warshaw, D.M., et al.: Differential labeling of myosin v heads with quantum dots allows direct visualization of hand-over-hand processivity. Biophys. J. **88**, L30–L32 (2005)
10. Sellers, J.R., Veigel, C.: Walking with myosin V. Curr. Opin. Cell Biol. **18**, 68–73 (2006)
11. Hammer III, J.A., Sellers, J.R.: Walking to work: roles for class V myosins as cargo transporters. Nat. Rev. Mol. Cell Biol. **13**, 13–26 (2011)
12. De La Cruz, E.M., Wells, A.L., Rosenfeld, S.S., Ostap, E.M., Sweeney, H.L.: The kinetic mechanism of myosin V. Proc. Natl. Acad. Sci. USA 13726–13731 (1999)
13. De La Cruz, E.M., Ostap, E.M.: Relating biochemistry and function in the myosin superfamily. Curr. Opin. Cell Biol. **16**, 61–67 (2004)

14. Rosenfeld, S.S., Sweeney, H.L.: A model of myosin V processivity. J. Biol. Chem. **279**, 40100–40111 (2004)
15. Sakamoto, T., Webb, M.R., Forgacs, E., White, H.D., Sellers, J.R.: Direct observation of the mechanochemical coupling in myosin Va during processive movement. Nature **455**, 128–132 (2008)
16. Forgacs, E., et al.: Kinetics of ADP dissociation from the trail and lead heads of actomyosin V following the power stroke. J. Biol. Chem. **283**, 766–773 (2008)
17. Coureux, P.D., et al.: A structural state of the myosin V motor without bound nucleotide. Nature **425**, 419–423 (2003)
18. Coureux, P.D., Sweeney, H.L., Houdusse, A.: Three myosin V structures delineate essential features of chemo-mechanical transduction. EMBO J. (2004)
19. Cecchini, M., Allosteric, A.H.M.K.: Communication in myosin V: from small conformational changes to large directed movements. PLoS Comput. Biol. **4**, e1000129 (2008)
20. Sweeney, H.L., Houdusse, A.: Structural and functional insights into the myosin motor mechanism. Annu. Rev. Biophys. **39**, 539–557 (2010)
21. Preller, M., Holmes, K.C.: The myosin start-of-power stroke state and how actin binding drives the power stroke. Cytoskeleton (Hoboken) **70**, 651–660 (2013)
22. Volkmann, N., et al.: The structural basis of myosin V processive movement as revealed by electron cryomicroscopy. Mol. Cell **19**, 595–605 (2005)
23. Veigel, C., Wang, F., Bartoo, M.L., Sellers, J.R., Molloy, J.E.: The gated gait of the processive molecular motor, myosin V. Nat. Cell Biol. **4**, 59–65 (2001)
24. Craig, E.M., Linke, H.: Mechanochemical model for myosin V. Proc. Natl. Acad. Sci. USA **106**, 18261–18266 (2009)
25. Hinczewski, M., Tehver, R., Thirumalai, D.: Design principles governing the motility of myosin V. Proc. Natl. Acad. Sci. USA **110**, E4059–E4068 (2013)
26. Dunn, A.R., Spudich, J.A.: Dynamics of the unbound head during myosin V processive translocation. Nat. Struct. Mol. Biol. **14**, 246–248 (2007)
27. Beausang, J.F., Shroder, D.Y., Nelson, P.C., Goldman, Y.E.: Tilting and wobble of myosin V by high-speed single-molecule polarized fluorescence microscopy. Biophys. J. **104**, 1263–1273 (2013)
28. Uemura, S., Higuchi, H., Olivares, A.O., De La Cruz, E.M., Ishiwata, S.: Mechanochemical coupling of two substeps in a single myosin V motor. Nat. Struct. Mol. Biol. **11**, 877–883 (2004)
29. Cappello, G., et al.: Myosin V stepping mechanism. Proc. Natl. Acad. Sci. USA **104**, 15328–15333 (2007)
30. Sellers, J.R., Veigel, C.: Direct observation of the myosin-Va power stroke and its reversal. Nat. Struct. Mol. Biol. **17**, 590–595 (2010)
31. Okada, T., et al.: The diffusive search mechanism of processive myosin class-V motor involves directional steps along actin subunits. Biochem. Biophys. Res. Commun. **354**, 379–384 (2007)
32. Shiroguchi, K., Kinosita Jr., K.: Myosin V walks by lever action and brownian motion. Science **316**, 1208–1212 (2007)
33. Karagiannis, P., Ishii, Y., Yanagida, T.: Molecular machines like myosin use randomness to behave predictably. Chem. Rev. **114**, 3318–3334 (2014)
34. Shiroguchi, K., et al.: Direct observation of the myosin Va recovery stroke that contributes to unidirectional stepping along actin. Plos Biol. **9**, e1001031 (2011)
35. Fujita, K., Iwaki, M., Iwane, A.H., Marcucci, L., Yanagida, T.: Switching of myosin-V motion between the lever-arm swing and brownian search-and-catch. Nat. Commun. **3**, 956 (2012)
36. Spudich, J.A., Watt, S.: The regulation of rabbit skeletal muscle contraction I. Biochemical studies of the interaction of the tropomyosin-troponin complex with actin and the proteolytic fragments of myosin. J. Biol. Chem. **246**, 4866–4871 (1971)
37. Wang, F., et al.: Effect of ADP and ionic strength on the kinetic and motile properties of recombinant mouse myosin V. J. Biol. Chem. **275**, 4329–4335 (2000)
38. Dunn, A.R., Spudich, J.A.: Single-molecule gold-nanoparticle tracking. Cold Spring Harb. Protoc. **2011**, 1498–1506 (2011)

39. Ortega Arroyo, J., et al.: Label-free, all-optical detection, imaging, and tracking of a single protein. Nano Lett. **14**, 2065–2070 (2014)
40. Komori, Y., Iwane, A.H., Yanagida, T.: Myosin-V makes two brownian 90° rotations per 36-nm step. Nat. Struct. Mol. Biol. **14**, 968–973 (2007)
41. Ohmachi, M., et al.: Fluorescence microscopy for simultaneous observation of 3D orientation and movement and its application to quantum rod-tagged myosin V. Proc. Natl. Acad. Sci. USA **109**, 5294–5298 (2012)
42. Nishikawa, S., et al.: Switch between large hand-over-hand and small inchworm-like steps in myosin VI. Cell **142**, 879–888 (2010)
43. Ries, J., Kaplan, C., Platonova, E., Eghlidi, H.M., Ewers, H.: A simple, versatile method for GFP-based single molecule localization microscopy. Biophys. J. **102**, 419A–419A (2012)
44. Kubitscheck, U., Kückmann, O., Kues, T., Peters, R.: Imaging and tracking of single GFP molecules in solution. Biophys. J. **78**, 2170–2179 (2000)
45. Hua, W., Chung, J., Gelles, J.: Distinguishing inchworm and hand-over-hand processive kinesin movement by neck rotation measurements. Science **295**, 844–848 (2002)
46. Blumberg, S., Gajraj, A., Pennington, M.W., Meiners, J.-C.C.: Three-dimensional characterization of tethered microspheres by total internal reflection fluorescence microscopy. Biophys. J. **89**, 1272–1281 (2005)
47. Knight, P.J., et al.: Two-headed binding of a processive myosin to F-actin: abstract: nature. Nature **405**, 804–807 (2000)
48. Burgess, S.A., et al.: The prepower stroke conformation of myosin V. J. Cell. Biol. **159**, 983–991 (2002)
49. Oke, O.A., et al.: Influence of lever structure on myosin 5a walking. Proc. Natl. Acad. Sci. USA **107**, 2509–2514 (2010)
50. Syed, S., Snyder, G.E., Franzini-Armstrong, C., Selvin, P.R., Goldman, Y.E.: Adaptability of myosin V studied by simultaneous detection of position and orientation. EMBO J. **25**, 1795–1803 (2006)
51. Enderlein, J., Toprak, E., Selvin, P.R.: Polarization effect on position accuracy of fluorophore localization. Opt. Express. **14**, 8111–8120 (2006)
52. Kodera, N., Yamamoto, D., Ishikawa, R., Ando, T.: Video imaging of walking myosin V by high-speed atomic force microscopy. Nature **468**, 72–76 (2010)
53. Ando, T., Uchihashi, T., Kodera, N.: High-speed AFM and applications to biomolecular systems. Annu. Rev. Biophys. **42**, 393–414 (2013)
54. Takagi, Y., et al.: Myosin-10 produces its power-stroke in two phases and moves processively along a single actin filament under low load. In: Proceedings of the National Academy of Sciences (2014)
55. Liu, A.P., Fletcher, D.A.: Actin polymerization serves as a membrane domain switch in model lipid bilayers. Biophys. J. (2006)
56. Howard, J.: The movement of kinesin along microtubules. Annu. Rev. Physiol. **58**, 703–729 (1996)
57. Schindler, T.D., Chen, L., Lebel, P., Nakamura, M., Bryant, Z.: Engineering myosins for long-range transport on actin filaments. Nat. Nanotechnol. **9**, 33–38 (2013)
58. Nakamura, M., et al.: Remote control of myosin and kinesin motors using light-activated gearshifting. Nat. Nanotechnol. **9**, 693–697 (2014)
59. Liber, M., Tomov, T.E., Tsukanov, R., Berger, Y., Nir, E.: A bipedal DNA motor that travels back and forth between two DNA origami tiles. Small **11**, 568–575 (2015)
60. Kalinin, S., Valeri, A., Antonik, M., Felekyan, S., Seidel, C.A.M.: Detection of structural dynamics by FRET: a photon distribution and fluorescence lifetime analysis of systems with multiple states. J. Phys. Chem. B **114**, 7983–7995 (2010)
61. Manley, S., et al.: High-density mapping of single-molecule trajectories with photoactivated localization microscopy. Nat. Methods **5**, 155–157 (2008)
62. Cognet, L., Leduc, C., Lounis, B.: Advances in live-cell single-particle tracking and dynamic super-resolution imaging. Curr. Opin. Chem. Biol. **20**, 78–85 (2014)

63. Krishnan, M., Mojarad, N.M., Kukura, P., Sandoghdar, V.: Geometry-induced electrostatic trapping of nanometric objects in a fluid. Nature **467**, 692–695 (2010)
64. Jin, S., Haggie, P.M., Verkman, A.S.: Single-particle tracking of membrane protein diffusion in a potential: simulation, detection, and application to confined diffusion of CFTR Cl channels. Biophys. J. **93**, 1079–1088 (2007)
65. Masson, J.-B., et al.: Mapping the energy and diffusion landscapes of membrane proteins at the cell surface using high-density single-molecule imaging and Bayesian inference: application to the multiscale dynamics of glycine receptors in the neuronal membrane. Biophys. J. **106**, 74–83 (2014)

Chapter 6
All Optical Label-Free Detection, Imaging and Tracking of Single Proteins

Parts of this chapter have been adapted from the following publication: Ortega Arroyo, J., Andrecka, J., Billington, N., Takagi, Y., Sellers, J. R. and Kukura, P. Label-Free, All-Optical Detection, Imaging, and Tracking of a Single Protein. *Nano Lett.* **14,** 2065-2070 (2014) [1] and are copyright (2014) by the American Chemical Society. All work presented in this chapter was performed by myself and Dr. Andrecka, with equal contribution in terms of experiments and analysis of each experiment. Dr. Yasuharu Takagi and Dr. James R. Sellers provided the myosin V constructs.

6.1 Introduction

Until now, the belief that the optical signal produced by an individual protein is orders of magnitude too small to be measured with an optical microscope has been considered a common dogma in optical spectroscopy. With the exception of fluorescent proteins, optical studies of proteins at the single-molecule level have depended on fluorescent labelling, but this has its own shortcomings as addressed in previous chapters. As a result, all-optical alternatives that achieve single-protein detection in the absence of any labels have been extensively sought.

An ideal label-free all-optical measurement for the life-sciences should satisfy the following criteria: not rely on the existence of a strong optical resonance, provide some level of information on the identity of the molecule, work under physiological conditions (temperature, presence of buffers, pH), be capable of providing information on dynamics, and possess imaging capabilities. Several alternatives have been proposed, however none of them fulfil each requirement. These alternatives have in common the need to either: amplify the weak signal from the molecule of interest, or control and subsequently suppress the background noise.

© Springer International Publishing AG 2018
J. Ortega Arroyo, *Investigation of Nanoscopic Dynamics and Potentials by Interferometric Scattering Microscopy*, Springer Theses,
https://doi.org/10.1007/978-3-319-77095-6_6

Under biological conditions, signal amplification can be achieved by either plasmonic—or cavity-based enhancement. All these previous approaches have in common that the protein signal is read out indirectly thanks to the presence of a structure with dimensions comparable to or larger than a protein.

Plasmonic-based approaches exploit the local surface plasmon resonance properties present in metallic nano-structures; specifically, the local electric field enhancement by several orders of magnitude over a very small volume otherwise known as a plasmonic hot spot and the local refractive index dependence on the plasmon resonance spectra, to indirectly detect signatures from single molecules. Examples using plasmonic hot spots have materialised in the form of surface-enhanced Raman scattering detection, which offers unique chemical fingerprinting and remarkable sensitivity [2, 3] at the cost of precise positioning of the molecule in the plasmonic hot-spot. Whereas, photo-thermal [4] and darkfield spectroscopy [5, 6] have capitalised on the detection of small spectral shifts caused by local refractive changes upon the binding of a molecule to the surface of a plasmonic structure. In both approaches, dynamics in the form of interaction kinetics can be extracted, but any spatial information of the single molecule event is unavailable by design.

The principle of cavity-enhancement relies on repeating the weak light-matter interaction multiple times. This is realised experimentally by coupling light into a micro-resonator structure, whereby a single-molecule event, in the form of substrate binding, increases the resonator optical path length and therefore, results in a shift of the cavity resonant wavelength. Despite being the most sensitive method available, [7, 8] the detected signal neither scales with molecular properties nor does it contain any spatial information, which complicates quantitative characterisation.

The other category of approaches, which read out the signal from a single protein directly, are based on noise suppression and more importantly strong electronic transition dipoles in the visible spectrum. Three approaches have materialised in the form of extinction [9], stimulated emission [10], and photo-thermal [11] detection. Despite the sensitivity, these approaches are unsuitable for studying dynamics in the context of biological systems due to the requirement of special imaging mediums, such as polymer matrices or glycerol, and the fact that most biological molecules lack strong optical resonances.

Yet another alternative for direct detection can be proposed on the basis that the size of a single protein is less than two orders of magnitude smaller than the diffraction limited area, which means that upon illumination, the amount of light scattered by the protein is non-negligible, constant and detectable. Therefore, the challenge in detecting the signal from a single protein, analogous to other direct detection approaches, consists in precisely measuring and removing the overwhelming background signal. In this chapter I provide a proof of principle study with the molecular motor myosin 5a heavy meromyosin, which demonstrates that interferometric scattering microscopy satisfies all the above requirements for a truly label-free all-optical measurement platform operating at the single-molecule level. The choice of myosin 5a is motivated by decades of single-molecule characterisation using different techniques and the fact that its processive properties serve as robust indicators of single-molecule detection [12].

6.2 Experimental Methods

6.2.1 Experimental Setup Parameters

For single-protein detection experiments, a $166\times$ magnification (63.6 nm/pixel) was used with a field of view of 104×104 pixel^2. Data was recorded at a frame rate of 1.7 kHz and the differential images time-averaged to 10 Hz. For single-protein tracking experiments with a higher sensitivity and bandwidth, the magnification was increased to $333\times$ with a field of view of 128×128 pixel^2. Data was acquired at 1.0 kHz and the differential images time-averaged to 25 Hz. The incident power for all measurements was adjusted to near-saturation of the CMOS camera for optimum sensitivity, which for a 1.0 ms camera integration time corresponds to an intensity of 2.5 and 10 kW/cm^2 for $166\times$ and $333\times$ magnification, respectively.

6.2.2 Sample Preparation

Rabbit skeletal muscle actin [13] and mouse myosin 5a HMM [14] with a C-terminal GFP were prepared as described. A 20 µM actin stock solution was prepared in polymerization buffer (10 mM imidazole, 50 mM KCl, 1 mM MgCl$_2$, 1 mM EGTA, pH 7.3 containing 2 mM DTT, 3 mM ATP). Actin was diluted in motility buffer (MB; 20 nM MOPS pH 7.3, 5 mM MgCl$_2$, 0.1 mM EGTA) 50–100 times.

The flow cell was rinsed with 1 mg ml^{-1} solution of poly(ethylene glycol)-poly-l-lysine (PEG-PLL) branch copolymer (Surface Solutions SuSoS, Switzerland) in PBS and incubated for 30 min. Next, it was washed twice with MB and actin solution was added. After 5 minutes of incubation the chamber was washed with MB and the surface was blocked by adding 1 mg ml^{-1} BSA in MB buffer and subsequent incubation for 5 minutes. 2–10 nM myosin solution (MB containing 40 mM KCl, 5 mM DTT, 0.1 mg ml^{-1} BSA and 5 µM calmodulin) was added, incubated for 5 minutes and then washed. Upon addition of ATP, myosin movement was observed.

6.3 Experimental Results

6.3.1 Label-Free Detection of Actin Filaments

Interferometric scattering was performed under confocal beam scanning illumination at $\lambda = 445$ nm to achieve higher levels of sensitivity compared to longer wavelengths. In the absence of myosin, the label-free imaging sample only consists of actin filaments immobilised on the glass coverslip by weak electrostatic interactions with the PEG-PLL branch co-polymer. Upon flat-fielding the raw images, individual

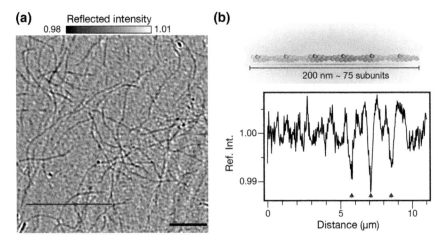

Fig. 6.1   Label-free detection of actin filaments. a Wide-field image of unlabelled actin filaments immobilised on a coverslip. Scale bar: 5 μM (black line). Marked cross-section: 11.5 μM (blue line). **b** Magnitude of the interferometric signal from individual actin filaments. Top panel: scheme illustrating the amount of material within a diffraction-limited area. Bottom panel: Corresponding cross-section from (**a**) with blue arrowheads marking the signal from three visible actin filaments

actin filaments together with nanometre glass roughness were observed as shown in Fig. 6.1a.

As described in Chap. 2, the magnitude of the interferometric signal depends on the scattering amplitude of the material found in within a diffraction limited area, which in turn scales linearly with the molecular weight of the object. In the case of an actin filament, a diffraction limited spot (200 nm) contains approximately 75 G-actin subunits which add up to a total molecular mass of 3.1 MDa (Fig. 6.1b). To estimate the relative contributions of the iSCAT signal produced by individual actin filaments and the remaining static background, line cuts were taken from each image (Fig. 6.1b). Here, actin filaments generated a signal on the order of 1.0 %, while the static background fluctuated by 0.3 %; thus ensuring a sufficient SNR to confidently distinguish the filaments. Comparison with previous experiments on another macromolecular complex, the simian virus 40 virus-like particles with a molecular mass of 15 MDa and an iSCAT signal of 4.5 %, [15] confirm the linear dependence of the scattering signal on molecular weight.

6.3.2   Label-Free Detection of Single Proteins

The linear dependence of the scattering signal on the number of protein molecules in the focus of the microscope poses the question whether individual proteins, specifically myosin 5, can be detected label-free. Based on the molecular weight of the specific myosin 5a construct used in this study (502 kDa), the iSCAT contrast for an individual protein is expected to be on the order of 0.15 %. We remark, that this construct contains two GFP moieties with strong optical resonance transitions (< 50 kDa), however the size of these moieties and choice of excitation wavelength would not mask the signal produced by the motor protein, as the expected iSCAT contrast from the GFP moieties would not exceed 0.01%.

Given the magnitude of the expected signal for myosin 5a, the static scattering background produced by individual actin filaments and the roughness of the glass overwhelms any contributions from the molecular motor. As a result, direct detection of myosin 5a is infeasible. However, upon addition of ATP myosin 5a HMM translocates along the actin filaments and thus becomes the only mobile scattering object in the sample. Hence, dynamic imaging (Chap. 3) should only reveal changes in sample scattering due to dynamic objects, in this case myosin 5a.

For dynamic imaging, an image containing all the static iSCAT features in a time-lapse video was generated (Fig. 6.2a). Next the number of frames in the acquired sequence that lacked a myosin 5a HMM signal was determined. This was achieved by subtracting the last frame from all other frames in the sequence and then time-averaging 20 consecutive differential frames together to reveal potential myosin 5a HMM signals. If the number of frames lacking a myosin HMM signal exceeded 100 then these images were averaged to produce the static iSCAT background image. Otherwise a temporal median filter was performed over the range of images in which the myosin 5a HMM molecule was processive, typically greater than 1000 frames, to avoid ghosting artefacts.

After subtraction by the image with the static features, the resulting frames contained signals due to mobile features and background noise. Under ideal conditions, the noise is dominated by fluctuations in the background level caused by shot noise in the detection of photoelectrons by the imaging system. Given that our camera saturates at 2×10^5 photoelectrons per pixel, the resulting baseline noise was on the order of 0.3 % root-mean-squared.

Even under ideal conditions, the background noise from a single shot masked the signal from mobile myosin 5a motors. We therefore averaged consecutive differential images to increase the detected photon count per pixel and with it the detection sensitivity. Specifically, we averaged 170 frames and thus reduced the detection bandwidth to 10 Hz, and as a result collected an equivalent of $\sim 2 \times 10^7$ photons per pixel. This translated into baseline fluctuations on the order of 0.024 %, thus enabling the detection of weak and mobile scattering features. Under these imaging conditions and in the presence of ATP, we observed several diffraction-limited spots that spatially overlapped with the actin filaments (Fig. 6.2b).

Furthermore, the distribution of iSCAT contrast was evaluated from 249 different diffraction-limited spots obtained from several tens of image stacks (Fig. 6.2c). No selection criteria other than colocalisation with an actin filament was considered in the analysis of the diffraction-limited spots. The contrast values were determined as the pixel value corresponding to the centre of mass of the point spread function, and only a single frame per diffraction-limited spot was included in the analysis. As a result, a uni-modal distribution with an average contrast of 0.18 % and a spread of about 30% was obtained in good agreement with our theoretical prediction based on

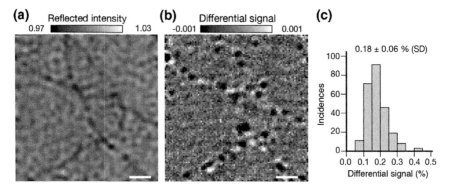

Fig. 6.2 Dynamic imaging of unlabelled myosin 5a. a Image containing purely stationary iSCAT features obtained by taking the median over a sequence of images of 500 frames. **b** Time-averaged dynamic iSCAT image after subtraction of the stationary features obtained from (**a**). Here 170 frames were binned together to improve the SNR, as a consequence there is an order of magnitude decrease in the intensity-scale from (**a**) to (**b**). Scale bars: 1 μm (black line). Raw images were collected at a camera exposure set at 0.40 ms with a frame time of 0.58 ms, [ATP]= 5 μM. **c** Distribution of the interferometric signal from 249 processive molecules collected over 15 replicates

molecular weight alone (0.15 %) and in correspondence to typical single-molecule intensity distributions. [16]

With regards to the spread of contrasts, this can be assigned to small variations in focusing, the lack of control over the orientation of the protein relative to the incident polarisation, and possible displacements of the protein relative to the substrate.

6.3.3   Detection Sensitivity of iSCAT

To characterise the effect of time-averaging on the detection sensitivity the standard deviation of the entire image at 166× magnification as a function of the number of frames averaged was calculated and compared with the visibility of the signal assigned to a myosin 5 motor. Qualitatively this is illustrated in Fig. 6.3a, where the cross-section of a differential image containing a signal assigned to a myosin 5 motor is displayed at different values of frame-averaging. For no frame averaging, the standard deviation amounts to 0.3 % as expected from the well depth of the imaging camera. However, as frame-averaging increases, the baseline fluctuations drop below the level of 0.2 %, thus enabling detection of the weak mobile scatterer.

The evolution of the image noise as a function of the number of averaged images follows shot noise behaviour up to 100 frames (Fig. 6.3b). Assuming that the 0.18% diffraction-limited spot signals correspond to individual myosin 5a molecules and that all proteins exhibit similar refractive index, we translated the image noise axis into the detectable molecular weight of a protein complex at a signal-to-noise ratio of one. Under this assumption the current experiment achieved a shot-noise detection

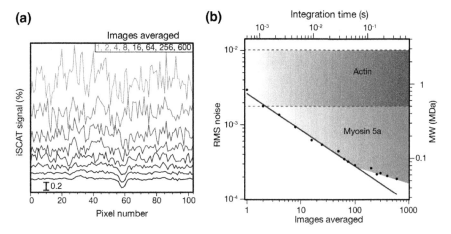

Fig. 6.3 Detection sensitivity of an iSCAT biosensor. a Horizontal cross-section of a diffraction-limited spot detected in dynamic imaging mode as a function of images averaged. The signal is assigned to myosin 5a due to its processive nature, characteristic 37 nm steps and contrast value of 0.18%. **b** Background noise in the biosensor as a function of the number of frames averaged and the corresponding camera integration time used in this measurement. Solid line indicates theoretical shot noise behaviour. Second vertical axis refers to the detectable molecular weight of a protein complex at a SNR = 1 using the signal from myosin 5a as a calibration point and assuming similar refractive indices. Dashed lines represent the average signal from individual actin filaments (blue) and myosin 5a (red). Shaded areas indicate denote regions where a SNR > 1 can be achieved

limit of 60 kDa, which corresponds approximately to a single bovine-serum-albumin molecule. After this point other noise sources, such as mechanical drift in the sample on the order of tens of nanometres and small fluctuations in laser intensity, dominate and no significant reduction in noise is obtained by temporal-averaging.

Furthermore, as long as the experiment is performed within shot-noise-limited conditions, the accumulation of photon counts per pixel can be performed either by spatial or temporal averaging. In fact, a background-subtracted image sequence taken 1.0 kHz at 333× magnification and then spatially binned to 166×, achieves the same sensitivity level as averaging four images together; thus resulting an overall gain in detection bandwidth. Thus, in principle all-optical detection of myosin 5a molecules is possible at frame rates approaching 1.0 kHz in our experimental setup.

6.3.4 Comparison of Single-Molecule Fluorescence and iSCAT Imaging

The specific binding to actin, the processive motion along these filaments in the presence of ATP, and the uni-modal iSCAT signal distribution suggests that the detected diffraction-limited spots indeed correspond to myosin 5a molecules. Nonetheless, whether these signals arise from aggregates rather than individual motors needs to

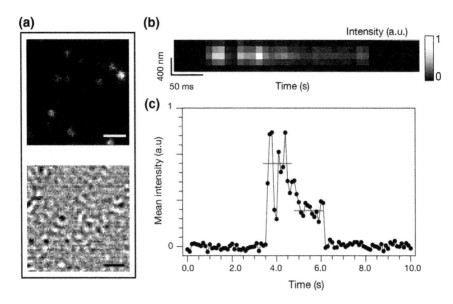

Fig. 6.4   Consecutive imaging of myosin 5a by TIRF and iSCAT. a Representative images of the same myosin 5a sample viewed by total-internal fluorescence (top) and interferometric scattering (bottom) microscopy. Scale bars: 1 μM. **b, c** Kymograph and fluorescence intensity time-series depicting the expected two-step photo-bleaching of the GFP fusion moiety found on a single myosin 5a molecule

be addressed. Aggregate formation would significantly alter the kinetics of translocation and in some cases may lead to the loss of function. Given that conventional single-molecule tests such as anti-bunching, single-step photo-bleaching and photo-blinking do not apply to iSCAT we performed single-molecule fluorescence assays using the same sample preparation by detecting the signal from the GFP fusion moiety (Fig. 6.4a).

The detected fluorescence signal followed the kinetics and two-step photobleaching expected from single myosin 5a molecules (Fig. 6.4b–c). Furthermore at the motor concentrations used, movement along the actin filaments was robust and no aggregates were detected. Upon transferring the same sample onto the iSCAT microscope, similar occurrence of translocation and transient binding events of myosin 5a were observed (Fig. 6.4a), suggesting that the signals in iSCAT are indeed representative of single myosin 5a molecules.

This experiment was performed in two separate microscopes and thus only consecutive rather than simultaneous iSCAT and fluorescence measurements were available as a consequence of the excitation wavelength used and the higher intensities required for iSCAT detection. However, correlative iSCAT and fluorescence could be easily achieved by shifting the iSCAT illumination further into the red, $\lambda > 520$ nm, so that no overlap with the absorption spectra of GFP occurs.

6.3.5 Observation of Single-Molecule ATP-Dependent Kinetics

As an additional proof for single-molecule detection the processivity of the detected myosin 5a signals was measured as a function of the concentration of ATP and the results compared with those obtained with other single-molecule methods [17, 18]. The run lengths and velocity were determined from single-particle tracking experiments, whereby the centre of mass of the PSF was calculated from a fit to a 2D-Gaussian function. Only molecules with actin binding and unbinding events were used in the determination of the run length distribution; whereas all the detected molecules were used for the run velocity. Run velocities were evaluated as the total distance travelled over the full observation time.

Velocities (Fig. 6.5a) and run lengths (Fig. 6.5b) at saturating ATP concentrations, 1 mM, were typical for single-molecule studies of myosin 5a [17–20]. Furthermore from measurements of run velocities at different ATP concentrations (Fig. 6.5c), the ATP binding and ADP release rate constants were extracted after fitting the data to a kinetic model previously described [18]. The extracted rate constants are in excellent agreement with the single-molecule kinetic studies.

6.3.6 Nanometric Tracking of Individual Myosin Molecules

When the SNR ratio was high enough to perform sub-10 nm localisation precision at a detection bandwith of 25 Hz, characteristic 37 nm steps were observed (Fig. 6.6). The localisation precision was assessed by the error in the fit from the 2D Gaussian function to the PSF given that detection of immobilised myosin signals for more than a couple of data-points was not possible. Nevertheless, these traces were obtained at 10 μM ATP, and thus contain segments without any visible steps. This can be alleviated by reducing the stepping rate via tuning the concentration of ATP.

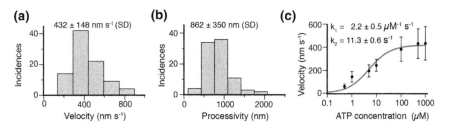

Fig. 6.5 **Processivity of unlabelled myosin 5a at the single-molecule level. a, b** Run velocity and processivity at saturating ATP concentrations (1 mM, n = 91). **c** Run velocity kinetics as a function of ATP concentration. The solid curve represents the best fit to the kinetic model: $V = ds/(1/k_1[ATP] + 1/k_2)$, where ds represents the average step size assumed to be 37 nm, k_1 the second order ATP binding rate constant and k_2 the first order ADP release rate constant

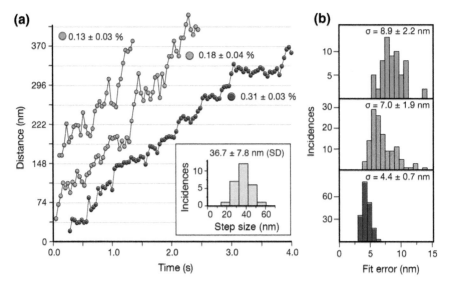

Fig. 6.6 Label-free single-particle tracking of myosin 5a. a Distance travelled for single myosin 5a molecules with different contrasts at 10 µM ATP concentration. Imaging speed: 25 Hz. Inset: Step size distribution. **b** Corresponding localisation precision assessed by the error associated with the fit for the different molecules shown in (**a**)

6.4   Discussion

The above observations and the excellent match between expected and observed iSCAT contrast for a single myosin 5a HMM molecule together with the following arguments strongly suggest that the observed moving objects are single myosin 5a molecules and not aggregates or other species. The movement of the detected signals was very robust, which is inconsistent with aggregates given the negligible amount of aggregation observed in the sample preparation. If the moving signals represented only a few percent of the total myosin 5a present, then the actin filament would have to be saturated with moving, non-detectable single myosin 5a molecules. Nevertheless, if such would be true the fluorescence assays performed would have shown a large saturation.

   In the experiments reported here, the sensitivity limits were near the 10^{-4} level and could not be further improved because of sample drift and nanoscopic motion of the actin filament. Since all objects on the sample surface produce an iSCAT signal, small changes in sample position due to drift or nanoscopic motion of the objects assumed to be static produce dynamic signatures that can potentially be confused for individual binding events. For instance, a 10 nm displacement of an actin filament with an original iSCAT contrast of 1.0% produces a differential contrast on the order of 0.1%, which is comparable to that of a single myosin 5a motor. As a result, the subtraction of time-averaged background images eventually introduces unwanted signatures and thereby departs from the desired shot noise limited behaviour. We reduced these effects partially in the experiments reported here by feedback stabilisation of the objective-sample distance to within 10 nm, but the dominant noise source for this proof-of-principle experiment was the residual motion of actin filaments.

At the time, these experiments were performed on a microscope placed on passively stabilised optical table and the samples were not securely held on top of the specimen holder. The use of more advanced vibration isolation equipment common to AFM/optical tweezers [21] and full 3D stabilisation, using for example the cross-correlation method described in Chap. 3, would ensure shot noise-limited sensitivity at the 10^{-5} level, enabling label-free detection with high SNRs, 10, down to the few tens of kDa. [22] Alternatively a camera with a larger full-well capacity could achieve the same sensitivity within a single-shot, and thus would reach a lower sensitivity in the time-scale where drift dominates the noise.

The main advantage of this approach is that it is compatible with most of the technologies developed for surface plasmon resonance sensing, i.e. detection of binding/unbinding events to specifically functionalised surfaces. The substrate becomes even more simplistic as it only consists of a glass coverslide, rather than a gold surface. The sensitivity is higher, with current limitations set largely by the availability of optimised detectors rather than the approach itself. The signal magnitude scales linearly with molecular weight and is independent of the binding site. One potential drawback of this approach is that little information about the molecule can be extracted beyond the molecular weight and that any scatterer is visible leading to potentially high backgrounds when residual scattering cannot be controlled.

6.5 Conclusion and Outlook

Our results disprove the notion that the scattering cross-sections of single proteins are orders of magnitude too small to be detected in an optical microscope. The presented detection modality does not require any specific molecular properties, such as strong transition dipoles, nor does it depend on sophisticated methodologies to reduce laser intensity noise or nanoscopic amplification of the weak single molecule signal. Instead, the imaging camera performs noise reduction automatically through the accumulation of detected photoelectrons with time and no ultrastable laser sources or specific refractive index environments are necessary. Together with the possibility of combining iSCAT with single molecule fluorescence [15] and the potential for unlimited observation times due to a lack of photobleaching our results enable novel applications from bio-sensing to multidimensional tracking of single biomolecules.

Most if not all label-free optical microscopy have focused their efforts to the detection of biological material in the form of single proteins, or DNA sequences. Nevertheless, there is a no reason why label-free detection should be limited to such applications, in fact one often overlooked application is the study of phase transitions on the nanoscale. On the basis that different phases exhibit different refractive indices and therefore varying degrees of scattered light, the nanoscale dynamic associated with these events can be directly visualised if high enough sensitivity is achieved, which we have demonstrated in this chapter to be possible. In this respect and in the context of biological systems, label-free single-molecule imaging could be applied to study lipid and membrane associated processes such as vesicle fusion or nanoscopic phase separation [23]. Similarly, the ability to image single proteins without labels may enable dynamic studies of self-assembly phenomena such as amyloidogensis [24] or microtubule/actin filament dynamics [25].

References

1. Ortega Arroyo, J., et al.: Label-free, all-optical detection, imaging, and tracking of a single protein. Nano. Lett. **14**, 2065–2070 (2014)
2. Nie, S., Emory, S.R.: Probing single molecules and single nanoparticles by surface-enhanced Raman scattering. Science **275**, 1102–1106 (1997)
3. Thacker, V.V., et al.: DNA origami based assembly of gold nanoparticle dimers for surface-enhanced Raman scattering. Nat. Commun. **5**, 3448 (2014)
4. Zijlstra, P., Paulo, P.M.R., Orrit, M.: Optical detection of single non-absorbing molecules using the surface plasmon resonance of a gold nanorod. Nat. Nanotechnol. **7**, 379–382 (2012)
5. Ament, I., Prasad, J., Henkel, A., Schmachtel, S., Sönnichsen, C.: Single unlabeled protein detection on individual plasmonic nanoparticles. Nano. Lett. **12**, 1092–1095 (2012)
6. Beuwer, M.A., Prins, M.W.J., Zijlstra, P.: Stochastic protein interactions monitored by hundreds of single-molecule plasmonic biosensors. Nano. Lett. **15**, 3507–3511 (2015)
7. Dantham, V.R., et al.: Label-free detection of single protein using a nanoplasmonic-photonic hybrid microcavity. Nano. Lett. **13**, 3347–3351 (2013)
8. Baaske, M.D., Foreman, M.R., Vollmer, F.: Single-molecule nucleic acid interactions monitored on a label-free microcavity biosensor platform. Nat. Nanotechnol. **9**, 933–939 (2014)
9. Kukura, P., Celebrano, M., Renn, A., Sandoghdar, V.: Single-molecule sensitivity in optical absorption at room temperature. J. Phys. Chem. Lett. **1**, 3323–3327 (2010)
10. Chong, S., Min, W., Xie, X.S.: Ground-state depletion microscopy: detection sensitivity of single-molecule optical absorption at room temperature. J. Phys. Chem. Lett. **1**, 3316–3322 (2010)
11. Gaiduk, A., Yorulmaz, M., Ruijgrok, P.V., Orrit, M.: Room-temperature detection of a single molecule's absorption by photothermal contrast. Science **330**, 353–356 (2010)
12. Sellers, J.R., Veigel, C.: Walking with myosin V. Curr. Opin. Cell. Biol. **18**, 68–73 (2006)
13. Spudich, J.A., Watt, S.: The regulation of rabbit skeletal muscle contraction I. Biochemical studies of the interaction of the tropomyosin-troponin complex with actin and the proteolytic fragments of myosin. J. Biol. Chem. **246**, 4866–4871 (1971)
14. Wang, F., et al.: Effect of ADP and ionic strength on the kinetic and motile properties of recombinant mouse myosin V. J. Biol. Chem. **275**, 4329–4335 (2000)
15. Kukura, P., et al.: High-speed nanoscopic tracking of the position and orientation of a single virus. Nat. Methods **6**, 923–927 (2009)
16. Kubitscheck, U., Kückmann, O., Kues, T., Peters, R.: Imaging and tracking of single GFP molecules in solution. Biophys. J. **78**, 2170–2179 (2000)
17. Yildiz, A., et al.: Myosin V walks hand-over-hand: single fluorophore imaging with 1.5-nm localization. Science **300**, 2061–2065 (2003)
18. Rief, M., et al.: Myosin-V stepping kinetics: a molecular model for processivity. Proc. Natl. Acad. Sci. USA **97**, 9482–9486 (2000)
19. Snyder, G.E., Sakamoto, T., Hammer III, J.A., Sellers, J.R., Selvin, P.R.: Nanometer localization of single green fluorescent proteins: evidence that myosin V walks hand-over-hand via telemark configuration. Biophys. J. **87**, 1776–1783 (2004)

20. Baker, J.E., et al.: Myosin V processivity: multiple kinetic pathways for head-to-head coordination. Proc. Natl. Acad. Sci. USA **101**, 5542–5546 (2004)
21. Perkins, T.T.: Ångström-precision optical traps and applications*. Annu. Rev Biophys **43**, 279–302 (2014)
22. Piliarik, M., Sandoghdar, V.: Direct optical sensing of single unlabelled proteins and super-resolution imaging of their binding sites. Nat. Commun. **5**, 4495 (2014)
23. Andrecka, J., Spillane, K.M., Ortega Arroyo, J., Kukura, P.: Direct observation and control of supported lipid bilayer formation with interferometric scattering microscopy. ACS Nano. **7**, 10662–10670 (2013)
24. Cohen, S.I.A., et al.: A molecular chaperone breaks the catalytic cycle that generates toxic $A\beta$ oligomers. Nat. Struct. Mol. Biol. **22**, 207–213 (2015)
25. Dumont, E.L.P., Do, C., Hess, H.F.: Molecular wear of microtubules propelled by surface-adhered kinesins. Nat. Nanotechnol. **10**, 166–169 (2015)

Chapter 7
Single-Molecule Chemical Dynamics: Direct Observation of Physical Autocatalysis

The work and data analysis routines presented in this chapter have been written by myself. The experimental work and acquisition of data was performed by myself and Andrew Bissette. The synthesis and subsequent ensemble characterisation of the reagents and products of the reaction including magnetic resonance, mass spectrometry and dynamic light scattering experiments were performed by Andrew Bissette.

7.1 Introduction

The field of life sciences has been one of the main targets for single-molecule studies, partly due to its synergistic development with optical microscopy methods, and the fact that the relevant length-scales are more closely associated with proteins rather than molecules. Nevertheless, there is no reason why these single-molecule investigations should be confined to a select number of disciplines. For instance, there are numerous applications in chemistry where ensemble measurements are not sufficient to measure, explain or characterise the observed dynamics, thus making them ideal targets for these type of studies.

One such case is autocatalytic reactions, whereby the product of a reaction acts as a catalyst. These set of reactions, specifically those involving the self-reproduction of lipid aggregates such as micelles and vesicles, play an important role from the origin of life perspective [1–3]. Namely because pure physical processes, such as component mixing upon the formation of self-assembled lipid aggregates, can increase the rate of product formation and thus drive the emergence of complex behaviour. Commonly classified as physical autocatalytic systems, these reactions are particularly attractive because they serve as models to investigate the dynamics of heterogeneous self-reproducing protocellular systems [4–6].

© Springer International Publishing AG 2018
J. Ortega Arroyo, *Investigation of Nanoscopic Dynamics and Potentials by Interferometric Scattering Microscopy*, Springer Theses,
https://doi.org/10.1007/978-3-319-77095-6_7

In terms of dynamics, physical autocatalytic systems are ideal for label-free single-molecule investigations, specifically because many of the observed phenomena remain poorly understood or have only been observed in the absence of any temporal characterisation [7, 8]. Vesicle growth, elongation, budding, and fusion are all examples of such complex behaviour that have been documented by using a combination of ensemble methods such as NMR spectroscopy [9], dynamic light scattering [7, 10], and FRET [11]. However, the major barrier to a comprehensive account of these processes has been the small size of the most relevant precursors: micelles and vesicles, which make direct and real-time detection currently not possible. Overcoming this limitation would provide valuable insight into the factors governing the behaviour of physical autocatalysts and the mechanisms by which micelles and vesicles reproduce. Developing a more complex model of these systems would, in turn, aid the development of novel protocellular systems, improve our understanding of compartmentalisation and self-reproduction of prebiotic systems, and on a more general level provide the technical framework to study assembly and phase transitions processes at the nanoscale.

In this chapter I present the application of iSCAT as a label-free sensor to monitor and characterise the progress of an physical autocatalytic chemical reaction in situ. The chemical reaction studied is an analogue of a recent example of physical autocatalysis, which is based on the conjugate addition of hydrophobic thiols to polar alkenes to give micelle-forming products [12].

7.2 Experimental Methods

7.2.1 Experimental Setup Parameters

For super-resolution and single micelle detection experiments, the magnification of the system was set to $333 \times$ with a field of view of 8.1×8.1 μm^2. iSCAT was performed under confocal beam scanning with a $\lambda = 445$ nm excitation. Data was acquired at 1.0 kHz and the differential images were time-averaged to 6.6–10 Hz bandwidth to improve the photon count. The incident power for all measurements was adjusted to 10 kW/cm^2 for 1 ms exposure time.

For reaction kinetic studies, the sample was monitored for at least 25 min at a sampling rate of 10 image sequences per minute, where each sequence contained 1000 images equivalent to 1 s of acquisition.

7.2.2 Sample Preparation

Borosilicate glass coverslips (No. 1.5, 24×50 mm, VWR) were first rinsed sequentially with Milli-Q water, ethanol and Milli-Q water. To increase hydrophilicity the coverslips were etched by bath-sonication in a 1 M HCl solution for 10 min. Then, the coverslips were rinsed extensively with Milli-Q water for 4 min and subsequently dried under a stream of dry nitrogen.

Reaction chambers were assembled by placing a CultureWell silicon gasket with a total holding volume of 15 μl (Grace Bio-Laboratories, Bend, OR) onto the glass substrate. The thiol-ene bond forming reaction was performed by: first adding 4 μl of Milli-Q water to the hydrophilic surface to create an aqueous layer, followed by 2 μl of neat, 9-cis-octadene thiol, thiol reagent; and finally 4 μl of 1.2 M of 2-methacryloxyloxyethyl phosphorylcholine (MPC) in 400 mM Cs_2CO_3 solution, MPC reagent. The thiol reagent, due to its hydrophobic nature, was pipetted gently onto the surface of the hydrophobic silicon gasket to increase the reaction surface area. Imaging started upon addition of the MPC reagent into the aqueous layer. All starting materials were filtered through 200 nm pores to minimise detection of extraneous scattering signals.

For the lateral interfaces and formation of small hydrophobic droplets of the thiol reagent, Milli-Q water and thiol were simultaneously pipetted onto different positions of the surface of the glass before the addition of MPC.

7.3 Results and Discussion

7.3.1 Detection of Micellar Aggregates as the Product of the Chemical Reaction

The system studied consisted of an autocatalytic reaction at the interface between an aqueous layer containing MPC and an organic layer containing the thiol reagent placed in an axial geometry on top of a microscope coverslip (Fig. 7.1a). The product of the reaction, a surfactant molecule, aggregates above a critical concentration and acts as a physical autocatalyst by increasing the surface area of interaction between the reagents. The surfactant aggregates diffusing into the aqueous layer were detected by iSCAT upon non-specific binding to the glass substrate.

To determine whether iSCAT has the sensitivity to detect individual surfactant aggregates, a solution containing exclusively these aggregates was added to the reaction chamber and subsequently imaged. The detection of these surfactant aggregates was achieved by the principle of differential imaging described in Chap. 3. Specifically, the set of acquired images was processed via differential imaging with a time offset $\Delta t = 100$ ms, temporally averaged to a detection bandwidth of 10 Hz, and finally 2×2 spatially binned to produce images containing diffraction-limited features on the order of 0.1% such as Fig. 7.1b.

It is important to emphasise that a running temporal average rather than just temporal averaging was applied for the analysis of all differential images, specifically the characterisation of the detected signals. By running a temporal average on the differential images, the rate of false positives was reduced and the recovery rate of

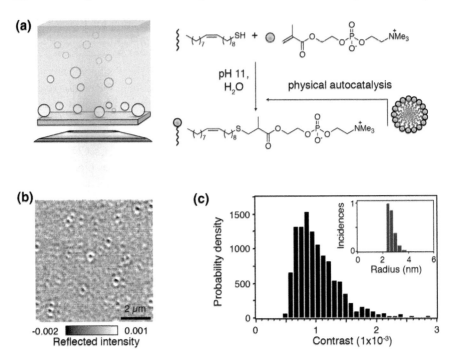

Fig. 7.1 Visualising physical autocatalysis by interferometric scattering microscopy. a Diagram of the biphasic reaction of aqueous MPC (orange) with neat water-insoluble 9-cis-octadene thiol (purple) carried out on a microscope coverslip. The product of the reaction, a surfactant, aggregates above a critical concentration and acts as a physical autocatalyst. Upon illuminating the sample with a coherent light source, surfactant aggregates (hollow spheres) bound to the surface, scatter light. Light that is backscattered from the surfactant aggregates together with that reflected at the aqueous/coverslip interface, contribute to an interferometric signal that is detected on a camera. **b** Representative image of single micelles binding to the coverslip surface after subtraction of the static scattering background. Scale bar: 2 µm **c** Distribution of the iSCAT contrast for a solution containing exclusively the surfactant aggregates. Inset: corresponding DLS number distribution

true positives was increased. This is a consequence that a single (un)binding event would be counted multiple times, contrary to a signal attributed to spurious noise.

To avoid repeated counts, single (un)binding events were only identified on the basis of having a trajectory length with at least four localisations and at most twice the length of the temporal average, in this case 100 images. Particle tracks, were generated by a modified cost matrix method described in Chap. 3. Assignments within the cost matrix were determined by the greedy approach; namely, by minimising the distance between features in consecutive frames found within a search-radius of 40 nm. Features with the minimum distance exceeding the search radius were classified as having no connectivity. The signal contrast associated to the binding events of this positive control exhibited a unimodal distribution with an average centred at 0.09% (Fig. 7.1c).

The average diameter of 6 nm for the surfactant aggregates as confirmed by DLS (Fig. 7.1c, inset), a typical iSCAT signal on the order of 0.1% for a single protein (Chap. 6), and the unimodal distribution of detected signal for the surfactant product, all provide evidence to strongly support that the diffraction-limited features correspond to individual micelles. Thus these results confirm that iSCAT has the sensitivity to image, localise and track individual micelles label-free.

7.3.2 Super-Resolution Imaging of the Progress of the Reaction

To monitor the progress of the reaction at the single-micelle and -vesicle level, data was acquired from an axial liquid/liquid interface. Here the underlying assumption is that to a first approximation, the rate of product binding to the surface is proportional to its concentration in solution, and therefore indicative of the progress of the reaction. The reaction occurred on a time-scale over tens of minutes and detection of the weak scattering signals required extensive temporal frame averaging, $\Delta t = 150$ ms. Therefore, continuous acquisition and subsequent data storage were infeasible with the available computational resources. Thus only a subset of time intervals, corresponding to one second of data every six seconds, were sampled leading to image sequences such as Fig. 7.2a.

Although general changes in intensity and local topography were observed in the flat-field images, no specific single-molecule signatures were discernible. However, using the same differential imaging and spatio-temporal averaging as in the positive control produced two types of diffraction-limited signals: dark and white, indicating particle binding and departure/fusion events, respectively (Fig. 7.2b). The differential images at the start of the reaction showed no binding events to the surface as expected from the absence of micelles in solution. After a lag-time of approximately ten minutes, particles landed on the surface at a rate that rapidly increased with time. A similar behaviour was observed for the unbinding events.

Imaging itself only provided a qualitative assessment of the progress of the reaction. By taking advantage of the stochasticity and relatively low binding density of events, additional information regarding the number and location of each scattering event was extracted by localising the centre of mass of each binding event and overlapping all of the positions onto a common map. As a result, super-resolution images of the position of each binding and unbinding/rupture event within the field-of-view of the microscope were generated (Fig. 7.2c–d). The apparent sparsity in the super-resolution maps was attributed to the non-optimal duty cycle of 0.16 caused by the limited sampling rate. Nevertheless this issue can be addressed in the future by either: increasing the sampling rate, using a higher full-well depth camera to minimise the amount of images averaged, or tuning the speed of the reaction.

In addition to providing spatial information of each binding event, these super-resolution images served to directly map dynamic interactions between the surfactant

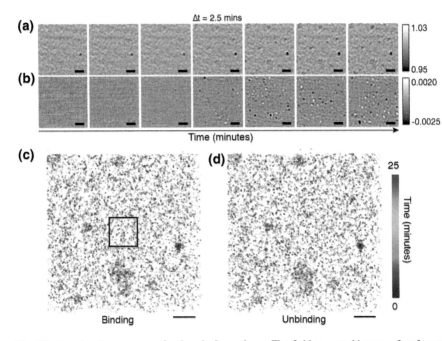

Fig. 7.2 Imaging the progress of a chemical reaction. a Flat-field corrected images of surfactant aggregates binding to the coverslip over time. **b** Corresponding differential images of (**a**) after temporal averaging of 150 consecutive frames and 2 × 2 spatial binning. The sign in the diffraction limited spots differentiate binding (dark) from unbinding/rupture (white) events. **c–d** Super-resolution maps identifying the centre of mass of each binding and unbinding event, respectively. The timing of each event is encoded with colour. Scale bars: 1 μm

aggregates and the substrate. In the case of Fig. 7.2c–d, despite the rather uniform distribution of events, areas with higher binding and unbinding frequency were observed, especially near the bottom middle portion of the images. These areas overlapped with the regions of smooth and slightly different intensity present in the flat-fielded images at later reaction times (Fig. 7.2a). Considering the linear dependence of iSCAT on the polarisability, and thus material properties of a nano-object, these specific areas likely correspond to a different substrate composition than the rest of the coverslip.

The super-resolution images for binding and unbinding also exhibited a significant level of spatial correlation. This could be interpreted as either a topographical rearrangement upon the interaction of the micelle with the substrate, or the subsequent unbinding of that micelle from an increasingly crowded surface. Further analysis of the timing between spatially correlated binding and unbinding events, and tuning of the functional groups on the surface could yield information regarding the underlying substrate-micelle and micelle-micelle interactions, and is a matter of ongoing investigation.

7.3.3 Direct Observation of Bilayer Formation

Upon imaging the reaction for longer times, on the scale of one or two hours, we directly observed the emergence of complex behaviour in the form of bilayer formation. This phenomena was characterised by a transformation in the relative roughness of the image, from local heterogeneity at the diffraction-limited level caused by individual surfactant binding, to completely homogeneous areas with significantly smaller fluctuations in signal contrast (Fig. 7.3). The onset of such behaviour correlated with the areas in the super-resolution image with higher binding and unbinding frequency as observed in Fig. 7.2c–d.

Although bilayer formation was partially catalysed by the incident illumination and the associated high intensities with confocal beam scanning iSCAT, the same phenomena was observed throughout the rest of the sample area, except with a delayed onset and much slower dynamics. This behaviour together with the frequent observation of diffusion on the newly formed interface are in agreement with similar experimental observations of macroscopic bilayer formation from vesicles deposited on a glass surface [13]. The main differences of the work presented in this chapter with the aforementioned iSCAT study are the chemical composition and size of the surfactant aggregates.

The observation of bilayer formation demonstrates the feasibility of generating complex structures and phase transitions from the autocatalytic reaction of simple prebiotic precursors partitioned across an aqueous/organic interface, a process required for the formation and reproduction of cell-like objects. Furthermore these results highlight the often neglected importance of surface interactions that can lead to complex behaviour. As previously mentioned, a thorough analysis of the super-resolution images between binding and unbinding events, together with substrate

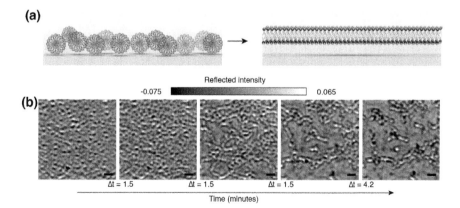

Fig. 7.3 Extended bilayer formation. a Cartoon illustrating the transition from nanoscopic micelles and vesicles to an extended bilayer structure on the coverslip. **b** Flat-fielded image time-series of the reaction between MPC and 9-cis-octene thiol after 20 min. Scale bars: 1 μm

modification would help elucidate the mechanism behind this nanoscale to mesoscale assembly process.

7.3.4 Characterisation of the Reaction Kinetics

The super-resolution images of the overall reaction provide information regarding the interaction between the surface and the product at the expense of information regarding the temporal evolution. However, this temporal information was recovered by analysing the corresponding time-course of binding and unbinding events (Fig. 7.4a). From these time-series the landing rates, i.e. the number of particles landed normalised per unit area and observation time, were determined as a function of reaction time (Fig. 7.4b). Under the assumption that the binding rates are a

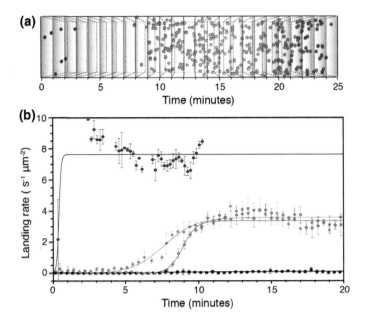

Fig. 7.4 Single-molecule kinetics of physical autocatalysis. a Super-resolution time-series emphasising the increase in binding events over time in the highlighted area of Fig. 7.2c. **b** Characterisation of the kinetics of physical autocatalysis by counting the number of binding events per unit time and area. Data points with error-bars represent the average and respective standard deviation of consecutive measurements sampled once every six seconds. Solid lines: fits for sigmoidal kinetics for the reaction between MPC and thiol, (blue/orange) and the reaction between MPC and thiol seeded with surfactant aggregates (purple). The seeded-reaction features an almost immediate high reaction rate upon addition of the surfactant aggregates. Solid black line and corresponding data points refer to the negative control consisting solely of the thiol starting material and an aqueous solution of Cs_2CO_3. All reactions were performed on the same coverslip

measure of the concentration of the product of the reaction, these landing rates provide a quantitative estimate of the reaction kinetics.

For the reaction between MPC and the thiol, the landing rates exhibited a lag time of 5–8 min followed by a short exponential increase in the landing rate. After 10 min the landing rate levelled-off analogous to a Langmuir adsorption isotherm (Fig. 7.4b, blue, orange). Although there are differences in the rate of increase between two similar reactions conditions, these could just reflect the spatial dependence of the observation area relative to the location of the reaction, in this case the liquid/liquid interface. Fits to a sigmoidal function provide evidence for the autocatalytic nature of the reaction; nevertheless neither the level nor the time at which saturation was reached reflect the completion of the reaction. Instead this observation can be attributed to the saturation of the surface, which is further confirmed by the almost identical unbinding kinetics.

A negative control, based on the absence of one of the starting materials, in this case the alkene reactant (MPC), showed no activity as expected (Fig. 7.4b, black). This suggests that the observed surfactant aggregates are neither thiol-in-water drops nor unwanted side-products of the reaction between the starting material and the silicone gasket or glass surface. By contrast, inclusion of 1 mM product in the aqueous solution eliminated the lag period, yielding rapid product formation upon addition of MPC (Fig. 7.4b, purple). It is not yet clear whether the difference in saturation levels is solely a substrate saturation artefact.

The observation of a lag period prior to an exponential growth in landing rate and its respective tuning under seeded conditions are indicative of a build up of the surfactant concentration. This concentration builds up until it reaches the critical micelle concentration, upon which aggregation and subsequent physical autocatalysis occurs. It is also of critical importance to mention that all these reaction kinetics were obtained from measurements on the same coverslips, thus validating the nature of the positive and negative controls and ruling out experimental artefacts such as cross-reactivity with other trace chemicals present in the sample. Taken together, all these experimental observations confirm that the reaction is indeed autocatalytic, and, on a more general level that iSCAT can be used to monitor a chemical reaction quantitatively at the single-molecule levels and completely label-free.

7.3.5   Observation of Physical Autocatalysis in Situ

In the previous experiments, the rate of product formation was correlated with the rate of binding events, given that direct imaging of the reaction interface was experimentally inaccessible. Namely, the liquid/liquid interface was located beyond the working distance of the objective, $\sim$170 μm. To address this issue and thus directly visualise the reaction in situ, the liquid/liquid interface was formed with a lateral configuration. This was achieved by pipetting the thiol solution directly on the coverslip followed by the aqueous solution, thus generating thiol-in-water droplets bound non-specifically to the glass surface (Fig. 7.5a). Under these conditions, various dynamic behaviours were observed.

Firstly for thiol droplets with radii on the order of a few μms, small surfactant aggregates budding directly out of the thiol-water interface and subsequently diffusing into the aqueous layer were observed (Fig. 7.5b). A fraction of these aggregates approached and non-specifically bound to the glass surface with autocatalytic binding kinetics analogous to the axial interface reaction conditions. In contrast to the axially arranged liquid/liquid interface experiments, the magnitude of the interferometric signal drastically varied over time. For instance, at the start of the experimental observation, the scattering signal from the aggregates was smaller than the constant scattering background present in a single frame. As the reaction proceeded, the intensity in the diffraction-limited features increased until reaching a magnitude similar to that of the neat thiol solution. This behaviour was consistent with the proposed physical autocatalytic mechanism; [12, 14] whereby the formed micelles and vesicles contain the thiol solution in their interior.

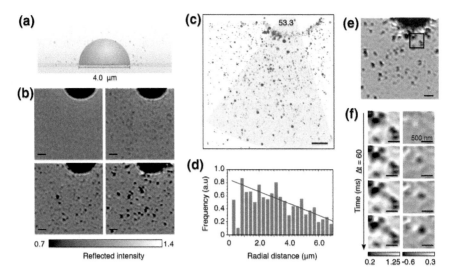

Fig. 7.5 **Observation of physical autocatalysis in situ. a** Illustration of the assay to achieve direct and in situ imaging of the reaction between MPC and a microscopic thiol droplet on the coverslip. **b** Representative flat-fielded images showing the progress of the reaction from left to right and then from top to bottom. The large-dark circular area on the top right corner corresponds to the thiol droplet. **c** Super-resolution map of the binding sites within the first seven minutes of the reaction. The colour and size of the plot markers encode the arrival time and signal intensity of binding events, respectively. **d** Dependence of the bound surfactant product density, found within the arc-sector depicted by the greyed-out region of (**c**), on the radial distance away from the droplet interface. Solid line: fit to a linear function. Scale bars: 1 μm. **e** Observation of the break down of the thiol droplet interface and appearance of extended lipid structures. Scale bar: 1 μm. **f** Flat-fielded and corresponding background-subtracted image time-series of the marked area in the (**e**). Scale bars: 500 nm

Secondly, the liquid/liquid interface was highly dynamic as characterised by increasing levels of local signal fluctuations (Fig. 7.5b). These fluctuations can be attributed to three main causes: the departure of products from the interface, which increase in frequency and size with time; the consumption of starting material; and the topographical re-arrangement of the droplet due to interactions with the substrate. Further analysis of the dynamics at the interface should allow the extraction and characterisation of parameters such as the bending rigidity and nanoscopic potentials involved in shaping the reactive pathways of physical autocatalysis; however they are beyond the scope of this thesis.

Thirdly, the distribution of the binding events exhibited a clear spatial dependence relative to the position of the droplet interface as evidenced by the super-resolution image of all binding sites within the first 7 min of the reaction (Fig. 7.5c). To quantify the dependence without bias, the density of binding aggregates per radial distance away from the interface was calculated as the number of events within an annular segment of the arc-sector defined by an angle of $53.3°$. Here, the centre of mass and size of the thiol droplet (4 μm) were estimated by solving a system of equations relating the lengths of two different chords, under the assumption that the surface cross-section of the droplet can be modelled as a circle. The resulting distribution showed a negative correlation between particle density and distance away from the reaction interface, as expected for a diffusion-limited sensing process (Fig. 7.5d). This result agreed with the observation of mostly weak signals in the axial liquid/liquid interface configuration, given that smaller particles posses a much higher diffusion coefficient and thus are more likely to bind earlier to the substrate compared to larger surfactant aggregates.

Finally, as the reaction proceeded beyond 10 min, the thiol-water interface almost entirely broke down and complex extended lipid structures erupting from the interface were observed (Fig. 7.5e). These structures proliferated rapidly on the 10 ms time-scale, and upon subtraction of the static and slowly varying scattering background, diffraction-limited sized events with increasing contrast were detected (Fig. 7.5f). Such events can be interpreted as the growth and possible division of individual vesicles from the reactive interface. In this case, such events were captured and imaged directly due to the presence of additional substrate interactions that confined the process to within the detection volume, given that this phenomena is likely to occur throughout the entire droplet interface during the course of the reaction. In the example highlighted, it was not possible to determine whether the newly-formed vesicle remained attached to the original vesicle or if division occurred. Nonetheless, this and other examples demonstrate that complex dynamics such as the growth and probable division of vesicles can be monitored in situ and label-free with interferometric scattering microscopy.

7.3.6 Interfacial Dynamics: Surface Interactions Lead to Different Mechanistic Pathways of Product Formation

For larger sized thiol-in-water droplets (radii > 100 μm), we observed a slow retreat of the thiol/water interface, which is consistent with product formation (Fig. 7.6a,b). Examination of such processes by differential high-speed imaging, at 1000 frames/s, revealed discretised behaviour. Namely, the interface retreats stochastically in nanometre-scale increments as opposed to a continuous process. Further overlay of the images confirmed that this quantised behaviour overlaps with the interface (Fig. 7.6b). The origin and underlying mechanisms governing such behaviour can be attributed to the formation of supramolecular aggregates on a vesicle-by-

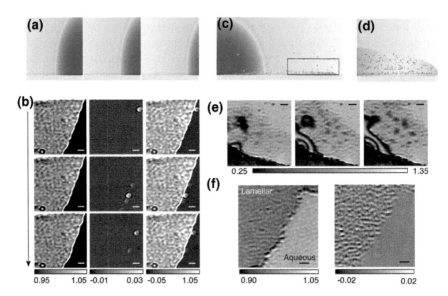

Fig. 7.6 Surface interactions lead to different mechanistic pathways of product formation. a Cartoon depicting the geometry of the assay where the thiol-water interface advances relative to the aqueous solution. **b** Flat-field corrected, differential and overlayed images of the advance of the thiol-water interface from left to right. Differential imaging reveals discrete events in the form of bright diffraction limited spots, while the composite image localises them to the interface. **c** Schematic illustrating the observation of an additional lamellar phase (grey) between the thiol (purple) and aqueous (orange) phases. **d** Zoom-in of the region marked (**c**) depicting that the lamellar phase is composed of lipid aggregates. **e** Corresponding time-series images showing the spread of such an intermediate phase represented by the highly heterogenous intensity distribution towards the middle portion of the frames. **f** Dynamics of the lamellar phase. Left panel: static image of the lamellar and aqueous phase on a glass substrate. Right panel: same image after background subtraction reveals high activity for the intermediate phase and none for the aqueous phase. Scale bars: 1 μm

vesicle basis directly at the interface. As a corollary, this confirms the ability of iSCAT to directly visualise product formation at the interface.

The main difference between the type of product formed, in this case vesicles due to their much larger contrast signal, as opposed to individual micelles can be explained by the presence of a solid substrate. Here, the solid surface introduces interactions and differences in local concentration that are absent in a liquid/liquid interface found in solution, which in turn bias the formation of vesicles (Fig. 7.6c, d). As a more dramatic case, in some occasions, a third highly dynamic phase at the interface between thiol and water was observed (Fig. 7.6e). This third phase, composed of different refractive index material as evidenced by the reflected intensity values, actively spread towards the aqueous layer. Furthermore, confined motion of multiple weakly scattering objects within this third phase, termed as lamellar phase, resulted in high local heterogeneity in the scattering intensity, which when viewed in a static picture appeared as roughness (Fig. 7.6f). Given the highly dynamic nature of the lamellar phase and the fact that the autocatalytic reaction still takes place, we propose that this phase is composed of a highly concentrated solution of surfactant aggregates products. These set of observations demonstrate that different mechanistic pathways exist according to the different spatial regions of the mixture, all of which can imaged directly and in real time with this interferometric technique.

7.3.7 Complex Phenomena in the Oil Phase

As there are no experimental limitations regarding the choice of imaging medium for this particular system, we performed measurements within the thiol phase and observed additional complex phenomena (Fig. 7.7a). For instance, at the edge of the interface, diffraction-limited water-in-oil droplets, likely signatures for reverse micelles, were visible. Initially these micelles, showed little to no motion. With time, however, these regions began to fuse into one another to form larger aqueous phases (Fig. 7.7b,c). Furthermore the interface of the larger micron-sized structures showed large intensity and positional fluctuations, suggesting a relatively low line tension [15].

Irrespective of the nature of the dynamics, the observation of diffraction-limited signals suggests the existence of reverse micelles. This makes the study of the reaction kinetics in both phases readily accessible, from which it would be possible to determine whether common features exist across both phases. Little information is currently known regarding the observed complex phase behaviour, thus making future investigations in this field a prime target for the next generation of label-free iSCAT.

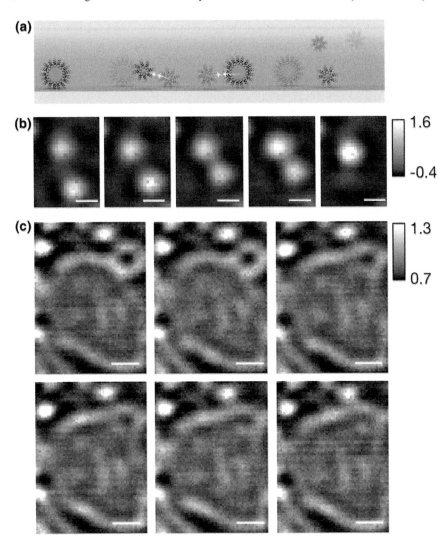

Fig. 7.7 Vesicle fusion events in the thiol phase. a Cartoon explaining the complex phenomena observed in the bulk thiol phase. **b** Fusion of small water-in-thiol droplets or reverse micelles captured in an image-time series. Scale bars: 500 nm. **c** Sequence of images illustrating the fusion of a small water-in-thiol or reverse micelle with a large water-in-oil droplet. Scale bars: 1 μm

7.4   Conclusion and Outlook

By applying iSCAT to the study of physical autocatalysis we have directly observed vesicle dynamics on the nanometre scale in situ and demonstrated the feasibility in performing quantitative kinetic studies with minimal amounts of starting material. In addition we have identified a diverse range of complex behaviours associated with interfacial dynamics and self-assembly at the nanoscale which are relevant to the origins of life [1, 3, 5]. While full elucidation of these processes is beyond the scope of this chapter, these phenomena clearly reveal the sheer complexity of these biphasic autocatalytic reactions and, on a more general level, reflect the relatively unexplored nature of this field.

Looking forward, as mentioned in the previous chapter, future studies in this direction will benefit extensively from the integration of tools and concepts from the fields of plasmonics and microfluidics and transform iSCAT into a useful benchmark method that complements existing ensemble techniques. It is just a matter of time and development of the sample area, before the same principles presented in this chapter are extended to similar systems such as supramolecular polymers, emulsions, aqueous biphasic reactions and liquid/liquid interfaces. Finally, this chapter has demonstrated that all-optical single-molecule approaches, especially those with label-free capabilities still have a bright future ahead of them, especially in fields beyond the life sciences.

References

1. Bissette, A.J., Fletcher, S.P.: Mechanisms of autocatalysis. Angew. Chem. Int. Ed. **52**, 12800–12826 (2013)
2. Ruiz-Mirazo, K., Briones, C., de la Escosura, A.e.s.: Prebiotic systems chemistry: new perspectives for the origins of life. Chem. Rev. **114**, 285–366 (2013)
3. Stano, P., Luisi, P.L.: Achievements and open questions in the self-reproduction of vesicles and synthetic minimal cells. Chem. Commun. **46**, 3639–3653 (2010)
4. Szostak, J.W., Bartel, D.P., Luisi, P.L.: Synthesizing life. Nature **409**, 387–390 (2001)
5. Segre, D., Ben-Eli, D., Deamer, D.W., Lancet, D.: The lipid world. Orig. Life Evol. Biosph. **31**, 119–145 (2001)
6. Walde, P.: Surfactant assemblies and their various possible roles for the origin(s) of life. Orig. Life Evol. B. **36**, 109–150 (2006)
7. Stano, P., Wehrli, E., Luisi, P.L.: Insights into the self-reproduction of oleate vesicles. J. Phys. Condens. Mat. **18**, S2231 (2006)
8. Berclaz, N., Müller, M., Walde, P., Luisi, P.L.: Growth and transformation of vesicles studied by ferritin labeling and cryotransmission electron microscopy. J. Phys. Chem. B **105**, 1056–1064 (2001)
9. Nguyen, R., Allouche, L., Buhler, E., Giuseppone, N.: Dynamic combinatorial evolution within self-replicating supramolecular assemblies. Angew. Chem. Int. Ed. **48**, 1093–1096 (2009)
10. Chungcharoenwattana, S., Ueno, M.: New vesicle formation upon oleate addition to preformed vesicles. Chem. Pharm. Bull. **53**, 260–262 (2005)
11. Chen, I.A., Szostak, J.W.: A kinetic study of the growth of fatty acid vesicles. Biophys. J. **87**, 988–998 (2004)
12. Bissette, A.J., Odell, B., Fletcher, S.P.: Physical autocatalysis driven by a bond-forming thiol-ene reaction. Nat. Commun. **5**, 4607 (2014)
13. Andrecka, J., Spillane, K.M., Ortega Arroyo, J., Kukura, P.: Direct observation and control of supported lipid bilayer formation with interferometric scattering microscopy. ACS Nano **7**, 10662–10670 (2013)
14. Buhse, T., Nagarajan, R., Lavabre, D., Micheau, J.C.: Phase-transfer model for the dynamics of "micellar autocatalysis". J. Phys. Chem. A **101**, 3910–3917 (1997)
15. Baumgart, T., Hess, S.T., Webb, W.W.: Imaging coexisting fluid domains in biomembrane models coupling curvature and line tension. Nature **425**, 821–824 (2003)

Chapter 8
Outlook

In this thesis I have presented the vast potential of iSCAT to study single-molecule dynamics at its three distinct domains: temporal resolution (Chap. 4), localisation precision (Chap. 5) and sensitivity (Chaps. 6 and 7). Here, the boundaries have been set by the available camera sensor technology and the design of the sample area, rather then by the technique itself as has been discussed in Chap. 2. As a result, future developments that push these boundaries will indubitably be intertwined with the arrival of new technology. Nevertheless, these advancements will only be incremental unless the underlying practical limitations of the technique are fully understood and adequately addressed.

In particular, possibly the greatest advantage of iSCAT over most single-molecule optical approaches, i.e. its label-free imaging ability, is also its largest limitation. The lack of specificity to differentiate nano-objects beyond the intensity of the signal and the characteristic dynamic behaviour of the object under study poses a considerable experimental challenge when dealing with complex multi-component systems. However, the lack of specificity can be partly alleviated by introducing a richer parameter-space, for example performing correlative fluorescence imaging, or by tailoring assays to the concept of differential imaging; for instance by introducing an off (unbound) and on (bound) state specifically to a target molecule by surface chemistry. In principle, it is feasible to study any sample with highly heterogeneous scattering backgrounds if: the background is time-invariant or varies at a much slower time-scale, the signal of interest moves in a time scale much faster than the background and the coherence length of the illumination source is sufficiently large for the interference term to occur.

Even if the aforementioned points are addressed, the experimental design merits consideration. From an instrumental perspective, the lack of full control over the sample stability, spurious reflections, and focus, constitute the main limitations for any experiment. These limitations effectively translate into non-shot noise limited measurements and, with it, the loss of decoupling of the temporal, localisation and sensitivity domains. From a sample perspective, high illumination intensities will

© Springer International Publishing AG 2018
J. Ortega Arroyo, *Investigation of Nanoscopic Dynamics and Potentials by Interferometric Scattering Microscopy*, Springer Theses,
https://doi.org/10.1007/978-3-319-77095-6_8

always pose a problem; especially for confocal beam scanning approaches, where the incident light is focused to a point on the sample. However upon using the concepts of recently developed lattice-like sheet microscopy [1] and total internally reflected illumination, the incident peak intensities can be dramatically reduced while keeping the average photon flux the same, thus lowering the sample damage without sacrificing signal to noise ratio. Furthermore all the results presented in this thesis were obtained with a CW laser as the illumination source, chosen solely on the basis of its low cost to output power ratio. However, the use of pulsed white-light source becomes not only attractive from the perspective of reducing unwanted interference with spurious reflections and allowing the system to relax (e.g. heat dissipation for AuNPs), but it also hails the opportunity to explore an additional parameter-space: the spectral domain.

At the moment, we can but glimpse at the many possibilities that lie ahead, primarily because the vast majority of single-molecule investigations and developments have been confined to a particular field. As I have shown in this thesis, specifically in Chap. 7, iSCAT microscopy is not bound to a single field (life sciences) but has many potential applications from fields as diverse as organic chemistry, nano-plasmonics, and condensed matter physics. As a result of the interaction with other disciplines, iSCAT will develop synergistically with them and exploit concepts common to those fields, for instance: microfluidics, plasmonics, surface patterning and surface chemistry, to mention a few. All in all, although iSCAT has developed to a stage where technology and the type of sample act as the current barrier, it is without doubt that widespread adoption of the technique as a powerful single-molecule tool will soon inspire a new generation of exciting developments and applications.

Reference

1. Chen, B.C., et al.: Lattice light-sheet microscopy: imaging molecules to embryos at high spa-tiotemporal resolution. Science **346**, 1257998 (2014)